# Process Management

Second Edition

Hans de Bruijn · Ernst ten Heuvelhof ·
Roel in 't Veld

# Process Management

## Why Project Management Fails in Complex Decision Making Processes

Second Edition

Springer

Prof. Dr. Hans de Bruijn
Delft University of Technology
Faculty of Technology, Policy
and Management
Jaffalaan 5
2628 BX Delft
Netherlands
e-mail: j.a.debruijn@tudelft.nl

Prof. Dr. Ernst ten Heuvelhof
Delft University of Technology
Faculty of Technology, Policy
and Management
Jaffalaan 5
2628 BX Delft
Netherlands
e-mail: e.f.tenheuvelhof@tudelft.nl

Prof. Dr. Roel in 't Veld
Waterbieskreek 40
2353 JH Leiderdorp
Netherlands
e-mail: roelintveld@hotmail.com

ISBN 978-3-642-42851-7 ISBN 978-3-642-13941-3 (eBook)

DOI 10.1007/978-3-642-13941-3

Springer Heidelberg Dordrecht London New York



Softcover re-print of the Hardcover 2nd edition 2010


*Cover design:* WMXDesign GmbH

Printed on acid-free paper

Springer is part of Springer Science+Business Media (www.springer.com)

# Table of Contents

# Chapter 1
# Introduction: Process and Content

## 1.1 Introduction

This book is about change. Change in complex issues. Change in complex issues always has the following three characteristics. Firstly, there are always multiple actors involved in this type of change. Put differently, the changes occur in a network of actors. These actors are, in a sense, dependent on each other. They need each other's support in effectuating the change, or, at the very least, they should be prepared not to frustrate or obstruct the change. The second characteristic is that these actors negotiate with each other. One needs the other, and vice versa. This is why the actors negotiate. The negotiations are complex in themselves. It is very unlikely that one negotiation session is sufficient to shape the change in a detailed and definitive way. More likely, this will require a series of meetings. And thirdly, the negotiation is a process, a series of meetings that can stretch across a longer period of time, sometimes even years.

During some periods, the negotiations may be intensive: there is a certain tension in the air, the actors meet frequently, and the process moves forward. At other times, however, things are relatively quiet. Summing up, this book is about negotiated change [9], which is shaped through a process.

This type of change processes takes place at various levels, ranging from the highest international level—negotiations between states, for instance about peace treaties and global environmental governance—to the micro-level of family members negotiating next year's holiday destination, and every possible level in between. No matter how unique all of these different levels are, they also have a number of common characteristics that result in strong similarities between the change and negotiation processes at all of these levels.

In recent years, the management world has become increasingly perceptive to the process-related aspects of change. Many changers define their own roles as those of process manager, mediator or facilitator. These qualifications indicate that they are careful not to have a hand in shaping the final result, and that they leave the content to others. It often becomes apparent that the substantive aspects of

H. de Bruijn et al., *Process Management*, DOI: 10.1007/978-3-642-13941-3_1,

change are complex to such a degree that for the change process to be successful, there is a need for explicit and specific attention to the process aspects.

This book analyses the process aspects of change. These aspects cannot be ignored when introducing a substantive change—a new organization, a new policy, a new technical system. The design of a change process, as well as its implementation and management, may have a major impact on the substantive outcomes. The substantive outcome of a change process depends partly on cognitive activities—analyses, calculations, model applications—but perhaps even more on the process that was followed. Who were involved in the negotiations? How did these actors play the game? Was there any resistance, how did this resistance come about, and what was done to overcome it? How much space was there for learning and for deviation from the original plans, which were the agreements for unforeseen circumstances, and so on.

This introductory chapter will start with a preliminary description of process management. Chapter 2 will contrast process management to several other management styles.

## 1.2 The Process Approach to Change: a Preliminary Description

Experiences with processes of change and management are often disappointing. The starting point of a change process is a substantive design (for instance a plan, a brief, a vision, a strategy, a technical design or a list of goals), but this is hardly recognizable at the end of the change process. The effectuated change may differ strongly from the desired change, or the change may not have happened at all.

What is the cause of these disappointing experiences? There are a number of possible explanations [5], which will be elaborated in more detail in Chap. 2.

- A first explanation is that the person who desires and initiates the change is often functioning in a *network of dependencies*. He depends on others, and can never impose one-sided change. After all, these other parties may amend, frustrate of even obstruct the change—during the decision-making phase as well as during the implementation.
- Secondly, many problems are so complex that there is *no unambiguous substantive solution*. As a result, the person who desires a certain change and supports this desire with substantive arguments often fails to convince the other parties. These parties may use other definitions of problems and solutions, and use their own substantive arguments. In other words, other parties have their own interests, which lead them to take a different view on the problem and to prefer different solutions. In cases when there is no hope for an unambiguous problem definition and solution, further research and analyses are often hardly productive: they only confirm the parties' notion of being right, or they increase the level of uncertainty.

- A third explanation is the fact that many changes are designed in a project-like way: strict problem definitions, clear goals, tight time schedules. However, a *project approach* has only a *limited meaning* in a network of dependencies. After all, the parties that the initiator of the change depends on will not simply accept the initiator's problem definition, goals and time schedules—so why would they cooperate with the planning of the change in question?

The essence is that an initiator of change depends on other parties, who may not be convinced by the initiator's substantive arguments. They may feel that their own ideas are not sufficiently reflected in the proposed change, and will therefore frustrate the project planning. Only when these other parties are involved in the change, they may recognize their own ideas in the problem definition and solution. And only then will they support the process. This illustrates the need for a process approach: the necessary involvement calls for a process of interaction between these parties. They need to discuss and negotiate the problems and solutions.[1] Once parties are aware of the fact that change can only be effectuated through a process of interaction and negotiation, they can make *process agreements*. These can be defined as:

### *1.2.1 Agreements About the Rules that the Parties will use to Reach a Decision*

These agreements about the rules usually precede the actual negotiation process. They are a part of the process of 'getting to the table' [15]. As has been noted before, it is not always easy to reach such agreements. After all, why would parties commit to a set of rules if they disagree, for instance, with the problem definition and solution proposed by the initiator of the change? This is even more relevant when the party in question is aware of the fact that the solution cannot be implemented without its cooperation. Parties will only be willing to reach agreement on the decision-making rules if these agreements offer them sufficient opportunity to serve their own interests. Rules should therefore always offer such a perspective to all parties in question.

In conclusion, a process approach to change implies that:

- the focus shifts from the content of the change to the way in which the content is developed and is implemented;
- there is prior agreement between the parties about the way in which the decision process is shaped;

[1] The notion of change management as process management can be found in several places in management literature: in the literature about change management in professional organizations (for example [1, 4]), about networks and network organizations (e.g. [5, 7, 10, 11]), about consensus building and mediating (e.g. [8, 13]), about stakeholder management (e.g. [6]), about management of change (e.g. [3]), and about shaping negotiation processes (e.g. [12, 14]).

- these process agreements offer each of the parties in question sufficient opportunity to serve their own interests.

It should be noted that while the process agreements may be explicit and fully formalized, their nature may also be informal.

Peace negotiations are usually characterized by formalized rules, down to the level of agreements on the set-up of the negotiation table and who will be seated where. Rules, however, are often used implicitly. When a manager aims to implement a minor change in his organization, he often follows a kind of process: he consults with certain persons, follows certain steps, has several back-up strategies in case there is no consensus, and so on.

### 1.2.2 The Process Manager and the Process Architect

As described above, the shift from a substantive approach to a process approach to change implies a shift in the role of the manager of the change. The manager is above all a *process manager*. He ensures that the process of change proceeds according to plan: the parties adhere to the rules, the parties are heard, communication is effective, decisions are made in accordance with the rules, and so on.

There is also a role to play for the *process architect.* In case of a substantive approach, the design is made by substantive experts. Examples include a policy official who designs a substantive policy plan based upon his expertise, or a spatial planner who makes a proposal for road construction based on an analysis of transport flows. In case of a process approach, on the other hand, there is a need for a process architect who oversees the realization of process agreements. It is the process architect who ensures that the process design is appealing to the parties involved: it should offer them sufficient opportunity to serve their own interests. In the case of the policy officer, these parties include those who will be affected by the policy, representatives of other, related policy areas, and potential financers of the new policy; in the case of the spatial planner, they include financers, provincial and local authorities, public transport companies, employer associations, environmental organizations, and so on.

This book describes a number of principles that may be useful in relation to the architecture and management of processes. These principles do not result in unambiguous recommendations about how to act under conditions a, b and c. Rather, these principles are notions that may be of importance, but that still leave room for a variety of processes. Put differently, there is not always one type of process that is preferable in a given situation. On the contrary: reality has shown that several types of process design and process managers may be effective. For instance, in an interesting analysis of two highly complex and sensitive processes, Curran et al. [2] demonstrate that various strategies may be effective. In addition, Holbrooke's 'tough' approach with regard to the conflict in Bosnia-Herzegovina and Mitchell's somewhat 'softer' approach regarding the conflict in Northern

**Table 1.1** Two effective process approaches to complex problems [2]

| | Differences between two process approaches | |
|---|---|---|
| | Holbrooke Bosnia-Herzegovina | Mitchell Northern Ireland |
| Fundamental objectives | Deal-oriented | Process-oriented |
| | Substantive | Relationship |
| | Transactional | Transformational |
| Fundamental interest/role | Mediator/advocate (with clout) | Mediator/neutral |
| Fundamental influence strategy | BATNA-focused | Joint gains-focused |
| | "Whatever it takes" | Model-future dealings |
| Coalition strategy | Simplify structure | Coalition of center against extremes |
| | Equalize core parties sequence | |
| Issue strategy | Process by fiat | Process an issue for negotiation |
| | Principles, then specifics | Procedure, then substance |
| | Sequential, the lock in gains | Separate into three strands, then package |
| | Defer deal-breakers | Decouple decommissioning |
| Process strategy | Engineer/manipulate representation | Principled inclusion (Mitchell Principles, "sufficient consensus") |
| | Sequence shuttles and Summits | Keep going at all costs ("variable geometry") |
| | Conceal/reveal information | Highly transparent |
| | Use Press to paint abyss/lower expectations, yet lock in gains | Relentlessly positive press spin |
| | Use process to build perceptions of personal credibility and power | Use process to build perceptions of fairness, dedication, and respect |
| Timing strategy | As fast as possible | As long as it takes |
| | Build 'momentum' by early wins, accelerating series of partial agreements, forcible actions, and process choices | Use time to bootstrap sense of respect, obligation, liking, and credibility |
| | | Expend personal credibility on final deadline |

Ireland both turned out to have advantages and disadvantages, but in the end they both proved effective. Table 1.1 shows the differences between the two approaches.

## 1.3 Structure of this Book

This book consists of three parts. Part I describes a number of introductory notions about process management. Chapter 2 contrasts process management with a number of other management styles. It addresses the main arguments in favour of a process approach, as well as the main risks.

Part II is about process design. Chapter 3 contains a number of principles that may be used in designing a process. Chapter 4 focuses on the actual design of a process: which are the key tasks of a process architect leading to a process design?

Once a process design is available, the process will need to be managed. Part III highlights such process management. Chapter 5 sets out how to guarantee the *openness* of a process. This openness concerns the actors to be involved and the issues to be placed on the process agenda. Chapter 6 discusses the *protection of the actors' core values*: how to prevent a process from infringing upon an actor's essential values, which is undesirable because it may very well result in this actor frustrating the process. Chapter 7 addresses the question of how to give a process *sufficient speed*. After all, processes can lead to extremely slow decision making. Chapter 8 discusses *content* in processes. How can a process be managed in a way that produces substantively rich results? How can we prevent the result being substantively poor for the sake of agreement?

The book concludes with a brief epilogue.

## References

1. Buchanan DU, Huczynsky A (1997) Organizational behavior: an introductory text. Prentice Hall, New York
2. Curran D, Sebenius JK, Watkins M (2004) Two paths to peace: contrasting George Mitchell in Northern Ireland with Richard Holbrooke in Bosnia-Herzegovina. Negotiat J 20(4):513–537
3. Dawson P (2003) Reshaping change: a processual perspective. Routledge, London
4. De Bruijn JA (2010) Managing professionals. Routledge, London
5. De Bruijn JA, ten Heuvelhof EF (2008) Management in networks, on multi-actor decision making. Routledge, London
6. Eden C, Ackermann F (1998) Making strategy. Sage Publications, London
7. Goldsmith SG, Eggers WD (2004) Governing by network, the new shape of the public sector. Brookings Institution Press, Washington, DC
8. Innes JE (1996) Planning through consensus building: a new view of the comprehensive planning ideal. J Am Policy Anal 62(4):460–472
9. Ives PM (2003) Negotiating global change: progressive multilateralism in trade in telecommunications talks. Int Negotiat 8(1):43–78
10. Klijn EH (2008) Complexity theory and public administration: what's new? Publ Manage Rev 10(3):299–317 (p 303)
11. Sabatier PA et al (2005) Swimming upstream, collaborative approaches to watershed management. MIT Press, Cambridge, MA
12. Sebenius JK (1991) Designing negotiations toward a new regime. The case of global warming. Int Secur 15(4):110–148
13. Susskind L, McKearnan S, Thomas-Larmer J (eds) (1999) The consensus building handbook. Sage Publications, Thousand Oaks
14. Watkins M (2003) Strategic simplification: toward a theory of modular design in negotiation. Int Negotiat 8(1):149–167
15. Watkins M, Lundberg K (1998) Getting to the table in Oslo: driving forces and channel factors. Negotiat J 14(2):115–135

# Part I
# Introduction to Process Design and Process Management

# Chapter 2
# Positioning the Process Approach

## 2.1 Introduction: From Deal to Process

How do these processes originate? How can we explain that issues that seem to be perfectly reconcilable with straightforward negotiation resulting in a clear deal, still develop into an unpredictable, seemingly never-ending process? The first reason is that in truly controversial cases it is impossible to start negotiating immediately. These issues have a past of negotiations and events that is so heavy with issues and failures that parties cannot simply rejoin the negotiation table. There is no longer any mutual trust, and trust cannot simply be restored by decree. Such negotiations therefore always have to be preceded by a process of 'pre-negotiations'. If these proceed well, they result in agreements about the 'real' negotiations. Pre-negotiations are highly contentious and are characterized by their own specific arrangements. They are usually carried out by 'unofficial representatives', they proceed via 'secret diplomacy', and may result in 'staged agreements' [33].

On 13 September, 1993, Israelis and Palestinians signed the Oslo peace accords. Formal negotiations had commenced in Norway on 11 June 1993. But the move 'to go to the table' was preceded by months of unofficial dialogue between the two sides. And even these unofficial dialogues could not simply be initiated. They were preceded by years of careful overtures. The problem in such processes is that groups that do not trust each other and sometimes do not even recognize each other need to talk to each other, and require mutual affirmation of the fact that the negotiations matter and that the negotiation partners have a certain degree of authority. But their official position is that the other party does not even exist—and as a result, affirmation of authority is a contradiction in terms that undermines one's own position. For how can any authority be assigned by a body that is not recognized and therefore has no authority itself?

For many years, so-called 'unofficial representatives' and 'entrepreneurial co-mediators' have made overtures towards each other. Unofficial representatives embody a critical combination of connections to important officials and unofficial status. They may for instance be authoritative academics who, under the veil of a scientific seminar, assess each others standpoints and test how far the other party is willing to go. Formally, these unofficial representatives have no governmental relationships. Both governments can easily dismiss any statements and concessions that the unofficial representatives

H. de Bruijn et al., *Process Management*, DOI: 10.1007/978-3-642-13941-3_2, 

make—although of course the governments can also take all the credit when it comes to potential successes.

Entrepreneurial co-mediators are 'moderate partisans' who reach out to moderate partisans on the other side. Often implicitly, they can build upon the work of the unofficial representatives. These overtures are highly contentious. If they became public, major unrest would immediately arise, and the negotiations would have to be stopped and even denied. At key moments in these negotiations, 'guardians' have to take control of the results. Guardians are top leaders who have established their credibility as protectors of their respective groups during crucial periods of danger and struggle. They possess the authority needed to gain widespread, grassroots support for the agreement. Rabin played this role on the Israeli side, and Arafat on the Palestinian side [33].

Pre-negotiation precedes the actual negotiation process. But even after the negotiation sensu stricto, a process can easily develop. How is that possible? In truly controversial processes, there is usually a complete lack of mutual trust. But even ordinary agreements require trust—if only the trust that the other party will conform to his part of the deal. What kind of agreements can be made in the absence of trust? The answer is: agreements that are the embodiment of a process themselves.

Watkins and Rosegrant [34] have analyzed a number of international negotiation processes, and have concluded that successful negotiations usually result in agreements about new processes, which in turn would be more focused on the implementation of the agreements [34, p. 63], for instance through:

- 'verification regimes arranging to observe each other's actions as a way of reducing mutual uncertainty and increasing transparency
- mutual deterrence making credible mutual commitments to devastating retaliation in the event of noncompliance
- incrementalism proceeding in a series of small and mutually verifiable steps, making future gains contingent on meeting current obligations, and embedding current negotiations in a larger context to avoid endgame effects
- hostage taking having each side deposit resources (such as a large sum of money) into an escrow account supervised by an independent party, with the understanding that the proceeds will be forfeited for noncompliance
- outside guarantors involving powerful external parties as guarantors of the agreement with the understanding that they will punish noncompliance.'

These are contentious and fragile pre-negotiations that precede the real negotiation process, which in turn is followed by processes in which mutually distrustful parties are seeking guarantees that the other party will adhere to the rules. This is how complex processes originate. Something that may enter the history books as an arrangement, a treaty between two parties that was concluded at a certain venue at a certain time, may in fact have a long history and a long future. The historical moment, represented as one point in time, hides a lengthy process.

A second reason why negotiations often turn into complex processes is that even the negotiations sensu stricto develop a process-like nature. It turns out that in a step-by-step process, parties are willing to make much more substantial concessions than in a single-leap process. This is what is sometimes referred to as 'progressive entanglement' [34, p. 223]:

Suppose the then US Secretary of State, James Baker, had gone to his Russian colleague Edward Shevarnadze right after Hussein's invasion of Kuwait to request Soviet support for

> a resolution authorizing a US-led multinational force to wage war on Iraq. Shevarnadze almost certainly would have refused. Nevertheless the Soviet Union supported such a resolution five months later, through progressive entanglement. First it was invited to support a joint resolution condemning Iraq's aggression, and then to support economic sanctions, next a defensive military operation, and ultimately the offensive military operation [34, p. 224].

A third reason why 'simple negotiations' can turn into complex processes is the fact that the negotiating parties of course have some notion of the functionality of arrangements such as pre-negotiations, post-negotiations and progressive entanglement. Initially such arrangements may have come about spontaneously, but in the second instance they are purposely introduced. This is how the notion of 'designing negotiations' was born [25]. This idea builds upon the assumption that a well-designed process is important in dispute management [2, 25, 32]. Analyses have been made of successful and less successful past negotiations that can generate lessons for future process design. One example is a comparison between the past strategy with regard to the ozone problem and a potential strategy to combat global warming [23]. Another is an analysis of the potential success of various general strategies relating to process design [25, p. 112]. In this regard, one of the key questions is: is it preferable to start by agreeing on a general framework of negotiated rules and principles, or should detailed agreement be reached on each individual issue—when the right moment presents itself—before the next issue can be addressed? Process-like arrangements have consistently proven to be crucial for success. Therefore these arrangements are purposely copied and introduced into ongoing processes. This reinforces the process approach.

## 2.2 Positioning Process Management

The term 'process management' is often used, and it may have several different meanings. In this chapter we will provide our view on the meaning of 'process management'. We will do this by contrasting the process with four other management styles [7]:

- process versus substance (Sect. 2.3),
- process versus command and control (Sect. 2.4),
- process versus project (Sect. 2.5) and
- process versus structure (Sect. 2.6).

In Sect. 2.7, this positioning results in an overview of the main arguments for process management. We will also set out potential different variants regarding the way parties support the results of a process (Sect. 2.8). We will then discuss a number of risks posed by process management (Sect. 2.9). This will help us to take a closer look at the process phenomenon (Sect. 2.10). We will conclude by briefly comparing the process approach with related approaches (Sect. 2.11).

## 2.3 Process Management Versus Substance

Process design and process management are the opposites of a substantive approach to decision making.

Many problems are unstructured by nature. We define unstructured problems as problems for which no unequivocal and/or authoritative solution is available. There are three possible reasons for this [10, 17]:

- no *information* is available that can be measured objectively;
- there is no consensus about the *criteria* to be used in solving the problem; and
- problems and solutions are dynamic.

To illustrate this, we will give an example of an unstructured problem that was solved with the help of a process design.

Urged by politicians, businesses intend to decrease the environmental impact of different types of packaging for consumer products. To do so, they have to identify the environmental harm caused by the various packages—such as the carton box, the glass bottle, the polycarbonate bottle and the polyethylene bag for packaging milk.

The environmental impact can be measured by comparing the various packages in relation to different environmental aspects, including toxicity, energy consumption, photochemical smog, emissions and waste.

Although the environmental impact might seem easy to determine, reality is different. The first problem is that objective information on environmental impact is hardly available. To measure a package's energy consumption, one must establish how much energy it takes to produce a package, to transport it to the consumer, to return it, if desired, and so on. Measuring these parameters objectively is hardly possible, if at all. Suppose, for instance, that the wood used for carton boxes is shipped from Latin America. This costs energy, but how much energy? Data is lacking, some of it is obsolete or is over- or under-aggregated. Methods to calculate a package's environmental impact are disputed and can never be fully objective. If the ship also transports washing machines and cars, how should the energy expenditure be divided amongst the wood, the washing machines and the cars? Furthermore, the problems' *system boundaries* are debatable: when a ship is needed to transport the wood, should the environmental impact of the production of the ship—and of other means of transport—also taken into account, for example?

Once information has been gathered about the environmental impact for each of the aspects, a type of package must be chosen. Which package has the lowest environmental impact? One that scores well on energy, poorly on emissions and neutral on waste? Or a package that scores poorly on energy, neutral on waste and well on emissions?

This is the second problem: there are no unequivocal criteria to compare the individual environmental aspects It is impossible to weigh such arguments neutrally and objectively—especially when environmental aspects are not only weighed against each other, but also against values such as economy and safety [9, 15, p. 23].

The package with the least environmental impact may simply be too expensive. There is also a tension between environment and safety. Baby food, for instance, may be packaged in glass jars that are reusable. The safety of such reusable jars is an issue, however: once cleaned, they may still contain glass fragments or residues of detergent.

It will come as no surprise, then, that different parties hold different views about the environmental impact of one and the same package. They have made their choices in terms of data, methods, system boundaries and the relative importance of environmental compartments. Each choice is debatable, so each outcome is debatable as well. Nor will it come as a surprise that each party has arguments to challenge the beliefs of the other parties [9].

A third characteristic of the substance of problems requiring a process design is the fact that they are *dynamic* [13]. The problem changes in the course of time. The logical consequence is that the answer to the question whether something is a solution for a problem also changes in the course of time.

Suppose the environmental impact of a package in the above example has been defined for the individual environmental aspects energy consumption, emissions, waste, et cetera). And suppose that there is consensus between companies, government and civil society about the criteria used (1) in the trade-offs between the various environmental aspects and (2) in the trade-offs between environment, economy and safety. Several developments may now take place that call the defined environmental profile into question. Technological innovations may be introduced. For example, it becomes technically possible to generate more energy from burning package waste than before. As a result, a package scoring poorly on energy may suddenly be found to score much better. Furthermore, producers of packages that scored poorly may make an effort to improve the environmental profile of their package, for instance by introducing energy-extensive production processes.

In the above examples, the information used earlier has become obsolete due to technological innovations. Criteria may also change. Suppose a country has a serious shortage of waste incinerators. This will cause waste reduction to become an important standard when assessing different packages. The opinion on the relative weight of criteria may change. Once the capacity shortage is alleviated, the relative importance of energy may increase again.

These dynamics may have a major impact on the way a problem is defined. The intelligent problem solver will find that the problem lies not so much in identifying the most environmentally friendly package, but rather in determining *how to sustain the process of continuous package improvement.*

If the problem is to use the most environmentally friendly package for milk, one of the milk packages will be chosen. If the problem is how to sustain the process of continuous package improvement, the solution might be that it is justified to maintain all four milk packages (carton, glass, polycarbonate and polyethylene), assuming that their mutual competition will minimize their environmental impact.

Dynamics then imply that attention shifts from finding the right problem definition and solution to a *continuous process* of formulating and solving problems. Any solution that was found today may be obsolete tomorrow.

### 2.3.1 From Objective to Authoritative: Negotiated Knowledge

Those who choose a substantive strategy of change regardless of the unstructured nature of problems will only create conflicts. After all, all choices are debatable. Instead, an initiator should accept the fact that different parties use different definitions of reality and may have good arguments for doing so. Even though a solution may never be assessed objectively, it may be *authoritative*, in which case it will be accepted by all parties involved. This requires involvement of these parties in the problem-formulating and problem-solving process. Problem definitions and solutions can be authoritative if they are the result of a process that has received input from the parties involved, in terms of their own information and values. During a process, the parties have negotiated about the data, system boundaries, methods and weighing criteria to be used. The result of such a process is what we call *negotiated knowledge.*

## 2.4 Process Management Versus Command and Control

Decision-making processes tend to take place in a network. Table 2.1 outlines the main characteristics of networks, compared with the characteristics of a hierarchy.[1]

When a decision-making process has to take place in a network, this always implies that several actors are involved in the decision making. They have different interests and are interdependent. No single actor can fully realize his own goals (interdependence). However, there are many differences between actors (pluriformity), which hampers cooperation and concerted decision making. In particular situations, particular actors may have no interest whatsoever in cooperating with others (closedness), which hampers decision making even further. Finally, the number of actors involved may change in the course of the decision-making process (dynamics): actors may join and leave.

In a network, hierarchical management stands little chance of success. A manager who wants to implement a project through command and control may seem decisive, but usually lacks the knowledge and power to implement his own views, and will therefore be met with considerable resistance in the network. Other parties can obstruct, delay or change his project. Hierarchical management may therefore be highly counterproductive: while the manager may appear to be decisive, he only creates resistance. The more hierarchy, the stronger the resistance in the network. Goals remain unachieved and plans unrealized.

**Table 2.1** Hierarchical and horizontal management

| Hierarchy | Network |
|---|---|
| Dependence on superior | Interdependence |
| Uniformity | Pluriformity |
| Openness | Closedness |
| Stability, predictability | Dynamic, unpredictability |

[1] Additional sources about networks and decision making: de Bruijn [8], Chisholm [3], Kenis and Schneider [20, p. 34] and Willke [35, p. 236].

The opposite of command and control is a process approach. As soon as a manager has to function in a network, he cannot simply rely on hierarchical managing mechanisms. After all, he depends on other parties, which will not automatically support him. A manager who recognizes this will not take unilateral decisions, but reach a decision in a process of consultation and negotiation with other parties. After all, such a process reflects the mutual dependencies in a network. This is what literature refers to as 'interest-based' decision making (e.g. [12]). It is worth noting here that an interest is regarded as a legitimate perspective on reality, rather than as an ordinary and one-sided perspective.

Thus, the packaging problems might seem a matter for business. After all, packaging is of strategic importance for the sale of consumer products. However, government and civil society, including the environmental movement, will also address these issues. Businesses partly depend on government and civil society. If the environmental movement feels, for instance, that businesses have chosen the wrong kind of packaging, it may seriously disrupt the corporate marketing. It may seek support from a government, which may apply its legal instruments. Moreover, there is not one single business approach. Perhaps the business in question has certain beliefs about the 'best' packaging, while other businesses in the chain—for instance the large supermarket chains—may hold other beliefs. And perhaps they hold so much power in the chain that their beliefs cannot simply be ignored. The conclusion is that when choosing their packaging, companies partly depend on other parties' support [9].

This does not mean, however, that there is no role whatsoever for command and control in networks. This management style may play a role in the process approach (see Chap. 7), but the dominant idea is that networks force those involved to adopt a process approach to decision making.

## 2.5 Process Management Versus Project Management

Thirdly, a process approach can also be positioned vis-à-vis a project approach. In a project approach it is assumed that problems and solutions are reasonably stable within certain limits. This allows for the use of project management techniques: a clear goal, a time schedule, a clear framework and a predefined end product. This will result in linear and structured decision making. Of course such an approach will only work in a static world. When a situation is dynamic rather than static, a project approach is impossible, and there is a need for a process approach.[2] This dynamic may have both external and internal causes.

External dynamics implies that an activity starts out as a project, but develops into a process because external parties, which introduce their own problem definitions and solutions, start to interfere with the project.

[2] This type of argumentation is common in literature about planning. See for instance Healey [16].

This is the familiar course of events in many infrastructural projects. Something starts out as a project (for instance the construction of a stretch of railway), and is met with resistance. A discussion arises, and parties try to obstruct or change the railway construction. After a while, the discussion is likely to be centred on issues that have little connection with the actual railway, for instance access to a certain area, the quality of living, and noise nuisance. As a result, the railway construction may lead to a variety of other issues being placed on the agenda. Something that starts out as a project thus ends up being a process.

Internal dynamics implies that an activity that starts out as a project develops into a process because the project owner realizes during the course of the project that the problem is different from what he anticipated.

A nice example is that of a house owner who decides one morning to move a certain painting. He then realizes that the colour of the wall behind the painting has faded, and decides to redecorate the entire wall, which affects other parts of the interior of the house, and eventually calls for a total refurbishment of the entire house. The next step is that he realizes that his renovation urge is related to the life phase he is going through, and he ends up in a psychologist's office. Something that started out as a simple project ends up being a complex process involving a range of other parties: other members of his household, a building contractor, neighbours, a psychologist, and so on.

Dynamics will be noticeable particularly when decision making has to take place in a network. After all, the various parties hold different views about how a problem and a solution should be defined.

Table 2.2 shows the differences between decision making in a hierarchy and in a network [5, 6, 11, 18, 19, 28, 30].

A hierarchy accommodates a decision-making process that is linear and structured, and that proceeds towards a solution via a number of different phases. The decision-making process is initiated by the actor who is superior in the hierarchy. The other parties participating in this decision-making process behave cooperatively, partly due to their subordination to the actor formulating the problem. Much of the decision making is a matter of project management.

In a network, there is no such project-like and phased development. Many problems may be formulated by one or more parties, but they never become the subject of decision making, or are never solved. The explanation is simple: the other parties see no sufficient reason to place the issue on the agenda, or they lose interest in the problem during the decision-making process. Parties may also notice

**Table 2.2** Decision making in a hierarchy and in a network

| Hierarchy | Network |
|---|---|
| Regular | Irregular |
| Phases | Rounds |
| Actors are stable, behave loyally and are involved in formulating the problem and choosing a solution | Actors join and leave, behave strategically; there are often winners and losers when the problem is formulated |
| Starting point and end clear | No isolated starting point and end |
| Problem → solution | Solution → problem |

during the decision-making process that their interests are harmed by the solution that seems to be available. As a result, they may try to obstruct further decision making. In other words, the parties participating in this managerial process behave strategically, partly due to the absence of a hierarchical subordination.

This may be illustrated by the decision making in relation to large infrastructural projects. Suppose the City of Rotterdam and the Rotterdam Port Authority intend to create additional industrial space in the port of Rotterdam through land reclamation. This allows for the construction of a new, large business area, called the Second Maasvlakte. According to the Rotterdam authorities, the underlying problem is the shortage of industrial space in the port area. It is not so difficult to imagine that a project-like approach has little chance to succeed. In order to reclaim land, Rotterdam depends on the support of other parties. Some of these will deny the fact that there is a lack of space, while others may acknowledge the problem but envision entirely different solutions—such as making more efficient use of existing space, rather than expanding. The stronger the City of Rotterdam's project-type steering and the stronger therefore the emphasis on its problem definition (lack of space) and problem solution (land reclamation), the smaller the chance of actual implementation. After all, for parties that do not identify with the problem, there is no incentive to support the City of Rotterdam. They will show either passive or active resistance, rendering the decision making random and unstructured rather than linear.

The idea of a regular and linear decision-making process should therefore be replaced with one of a process that takes place in *rounds* [28]. In a round, actors either reach a decision during a fight, or rather try to prevent doing so. A round will finish at one particular point in time and produce a provisional result, involving winners and losers. This might seem to conclude the decision making, but a new round may suddenly present itself.

This random course of decision making can be explained by the behaviour of the parties involved. The following examples illustrate behavioural patterns that are perfectly rational in a network, but that still result in the randomness mentioned before. We will use the land reclamation example as a reference in each case.

### 2.5.1 Dynamics

There may be new developments that call for a redefinition of problems or solutions. Parties that regard these new developments as a chance of influencing the decision in their favour will introduce these into the decision-making process.

In the Rotterdam area, the issue of 'recreation' may be high on the agenda: general opinion holds that there is a lack of recreational facilities. This may lead to the original problem being redefined: it includes not just the lack of space, but also the quality of life in the area. This redefinition may be opportune for the City of Rotterdam. Perhaps support for the Maasvlakte area could be swapped for Rotterdam's support for the expansion of recreational facilities. Or perhaps part of the reclaimed land could be designated as a new recreational area. In either case, the chances of implementation of the Maasvlakte area will increase. In this case, it may be clear that a project-like change will be little effective. Those who take a project-like approach will regard a changed problem definition as a problem rather than as a solution or an opportunity.

### 2.5.2 Compensation of Losers, and Coupling

There are always winners as well as losers when it comes to decision making. In a new round, losers may try to make up for their losses, and thus try to obstruct decisions that have been taken. This 'making up' may, however, also be a conscious act on the part of the winners: they may compensate the losers of decision-making process A in a following decision-making process B. The result is a coupling between A and B, which will seriously hamper a project-like approach to issue B.

One of the potential losers in the decision making regarding the Maasvlakte area is Hook of Holland, one of the smaller communities in the broader Rotterdam area. After all, the new industrial area may be constructed within sight of Hook of Holland. It is therefore not unthinkable that this community will be compensated for this, for instance by an increased number of 'bad weather facilities'. This will make Hook of Holland more attractive to tourists and daytrippers. Suppose the construction of a swimming pool is one of the compensatory measures. The result is a coupling between the decision making about land reclamation and that about a swimming pool in Hook of Holland. While this coupling to Hook of Holland's problems may help to implement the land-reclamation project, it also shows that a linear, project-type approach is hardly fruitful.

### 2.5.3 Solution Seeks Problem

When the parties have chosen a particular solution, they tend to look for problems that fit this solution. After all, proving that their own solution solves another party's problem may gain them this other party's support.

The construction of an airstrip in the Maasvlakte area may positively affect another sensitive issue in The Netherlands: the spatial problems regarding Amsterdam Airport Schiphol. The Maasvlakte solution would then be coupled to the Schiphol problem. For Rotterdam, this may be an attractive strategy, because it may result in the support of certain actors—perhaps even of the Minister of Transport, Public Works and Water Management, who is troubled by the Schiphol issue and is eager to find a solution. Again, this is an effective strategy. However, it cannot be understood from a project-managerial perspective, because it may make decision making more random. It suddenly involves not just lack of space and land reclamation, but also airport infrastructure.

### 2.5.4 Blocking Power Towards the End of the Decision-Making Process

A common phenomenon is the fact that actors start acting only towards the end of the decision-making process: they recognize which solutions are likely to be chosen, they disagree with these, and they attempt to obstruct or change any further decision making. To use the terminology of the rounds model: they enter the scene in the final rounds, and only then do they take part in the decision-making battle.

### 2.5.5 *Strategic Behaviour*

Finally, it should be mentioned that actors may show strategic behaviour. Those who know that decisions are taken in rounds can adapt their strategic behaviour accordingly. Actors may, for instance, decide to adopt a reserved attitude in round $x$ and accept a loss in order to have a stronger position in round $x + 1$. Alternatively, they may show much resistance in round $x$ and offer compensation in round $x + 1$.

Smaller communities may join in the land reclamation game. It may be little effective for them to resist the expansion of Rotterdam, but resistance may be interesting from a strategic point of view. Even though they are only minor players, the stronger their resistance, the better the chances of being compensated for losses in the future.

Each of these behavioural patterns shows that project management stands little chance of success. Project rationality may even be counter-productive because it fails to make efficient use of the opportunities presented by random and unstructured decision making.

The alternative is process rationality: an initiator acknowledges the fact that he depends on other parties, and invites these parties to join in a negotiation process in which these parties couple their problems and solutions. From the perspective of process rationality, such behavioural patterns are no aberrations, but perfectly normal. The behaviour of these parties will result in some unexpected twists and turns, but as long as the parties in the process engage in negotiation, these will cause little damage and may even prove useful.

## 2.6 Process Management Versus Structure

An important characteristic of networks and network-like organizations is pluriformity—an omnipresent theme. Youth care, for instance, is often criticized for being too fragmented: there are too many different authorities, and they sometimes operate in complete isolation. Decision making regarding large infrastructure is always criticized for involving too many parties, which results in sluggish decision making. The same complaint is heard within organizations: the organization has a patchwork-like structure, with too little cooperation between the various units.

Those who are repeatedly confronted with such fragmentation will be tempted to seek a solution based on structural measures. They establish a central authority, for instance, to manage the various organizations. Alternatively, competencies are redistributed, or mergers are planned. Such structural measures may be helpful, but they have a number of important limitations.

The following example will illustrate this:

Many of the world's metropolitan areas are faced with the issue of the right managerial level of scale. There may, for instance, be an issue relating to traffic and transport, and this issue cannot be solved at the level of individual cities. It may become apparent that there is no managerial authority at the level at which these problems can in fact be solved—for instance at the regional level. In that case it is easy to conclude that there is a need for a structural change. At the right level of scale, a managerial authority is to be installed that can tackle the traffic and transport issue decisively. There are at least two possible objections against this line of reasoning.

Firstly, the right managerial level cannot always be objectively determined. Suppose a major city is divided in two parts by a river with a tunnel underneath it. Suppose this tunnel has at least two functions: (1) it connects the northern and southern parts of the city; and (2) it is an important bottleneck in a motorway. Which is the right level of scale for decision making when it comes to upgrading the tunnel? From the perspective of function 1, the right level is that of the city council; from the perspective of function 2, it is that of the road authority. The city council may hold the opinion that the congestion surrounding the tunnel should be solved as soon as possible, while the road authority may hold entirely different views: the bottleneck has a positive function because it prevents congestion in other parts of the road network. The 'right level of scale' may seem to be a clear and unambiguous criterion, but in practice it is not.

Secondly, this is only one problem among the many other problems that this area undoubtedly has. Suppose we can identify twenty clusters of major problems—which is probably a conservative estimate. And suppose we design the right managerial level of scale for each of these problems. The result will be twenty 'right levels'. Let us make a generous assumption: half of these levels are identical to other levels—so the actual number of levels of scale is ten. The conclusion is obvious: the more issues, the less useful it is to argue in favour of the 'right level of scale'. A structural change that results in the optimal level of scale for issue A can result in a sub-optimal level of scale for issue B. Moreover, the 'right level of scale' is not a static concept. Problems change, perceptions of problems change, and therefore the notion of what constitutes an optimal level of scale also changes.

If structural solutions are ineffective, an alternative approach is to invest in better processes. We accept the fact that there is fragmentation, and try to improve the processes among the different units. How? By making process arrangements which set out mutual cooperation. Or by analyzing the major bottlenecks in the road network jointly—city councils, road managers, businesses together—and designing joint solutions. Attention shifts from structure design to process management.

## 2.7 The Main Arguments for Process Management

This perspective on process management results in a number of arguments for process management.

### 2.7.1 *Reducing Substantive Uncertainty*

Having all the relevant information available is crucial when it comes to solving unstructured problems. In many cases, the parties involved have different information, which is essential for the adequate solution of a problem. Testing the different sources of information against each other may improve the quality of the information used. In order for such a confrontation to take place, the relevant parties should be involved in the problem solution.

An adequate example to illustrate this is the construction of a space shuttle: a project that is technically highly complex. A major problem when it comes to such projects is *unruly technology*. The construction of various components requires the most advanced technical knowledge, which in many cases has not yet been sufficiently developed. The designs used still contain uncertainties that call for further research, or for an experiment. There are only limited possibilities to test the technical options chosen. The results of such tests tend to be open to more than one interpretation. In short, there is no 'hard' and objective knowledge available to solve the technical problems.

As a result, the management is faced with many uncertainties and incomplete information during the construction of the space shuttle. Nevertheless, choices have to be made. It is never certain whether or not the technical choice in question carries too great a risk.

In such a situation, an organization should strive to minimize the risk of these choices. Managers use jargon such as *debugging* (removing errors from the system) and *closure* (soothing a debate between professionals because consensus has been reached). Within the NASA organization, this is put into practice by a type of process management. Every risky decision is subjected to a formal *check and double-check* procedure: every choice made by the technical specialists is subjected to a process of counterchecks within the organization, during which the choice is screened.

These *review procedures* have been formalized. Any uncertainties that remain after such procedures are formulated as accurately as possible and presented to the management. The management will then either decide that a technical risk is justified, that money and time will be allocated for further research within the existing design, or that there is a need for a new technical design. Process management is thus an important tool for the management to uncover technical imperfections [31].

### *2.7.2 Enriching Problem Definitions and Solutions*

Different parties tend to have entirely different perceptions of and (normative) beliefs about problems and solutions. Testing these different views against each other may have an enriching effect [29]. If such a confrontation is to take place, the relevant parties have to be involved in solving the problem.

One argument for the process approach is of course support, as will be explained later. For many people, support has a negative connotation: it is a necessary evil that will compromise the quality of the decision making. As a consequence, they feel that good, substantive ideas are supposed to be watered down for the sake of gaining support. The argument of enrichment contests this line of reasoning. In this argumentation, support and substantive enrichment of ideas may well go hand in hand.

The board or the management of a professional organization wishing to develop a strategy for its organization (a strategy that is more than a paper strategy, i.e. a strategy that actually influences the organization's functioning) will be unable to do so without involving the professionals. Such involvement will promote support for a strategy, and tends to enrich it as well. After all, the professionals in the organization know the opportunities and threats in the organization's environment better than anyone else, and they also know the organization's potential to respond to them.

The argument of enrichment not only concerns the common product of a process (in the above example: a strategy). Enrichment may also relate to the knowledge of an individual party. Thanks to the process, a party may develop a richer view itself. After all, it learns about the perspectives and beliefs of the other parties. The process may also help it to develop a better understanding of and appreciation for the views of the other parties. Both types of enrichment can contribute to gaining support.

### 2.7.3 Incorporating Dynamics

As has been mentioned before, the chosen problem definitions and solutions can become obsolete fairly soon due to the phenomenon of dynamics. This allows unwilling parties the opportunity to distance themselves from a chosen solution, invoking new information, new solutions available, and so on.

In most cases, the challenge lies in preventing such new insights and information from being available outside, rather than within, the problem-solving process. The only way to achieve this is to involve all relevant parties in the solution-seeking process, since they are the carriers of new insights and information.

Business and civil society consult with each other about the environmental impact of packages. It may be important for particular experts or research institutes to join this process, because in many cases they are the carriers of innovations. If they do not participate in the process, they will publicize their new insights outside the process at some stage. This can lead to a situation in which parties within the process are consulting with each other about particular options, while parties outside the process have known for quite some time that innovations are forthcoming, making this discussion outdated. Involving these innovative parties ensures that the dynamics can be discussed during the process. This is often the reason why parties are willing to consult with highly critical opponents. They prefer this criticism to surface in the context of a process rather than it being made public outside of the process. The process at NASA outlined above makes a good comparison: criticism is organized, because it had better be discussed in the context of the process rather than being allowed to lead a life of its own within the organization (or worse: outside of it) [9].

When a process leads to such incorporation of dynamics, the parties are in a position to learn. After all, they are constantly faced with others and with new views, which can stimulate them to reflect on their own views. (Giddens [14] refers to this as 'dialogic democracy', which could lead to an increase in social reflexivity).

### 2.7.4 Transparency in Decision Making

Decision-making processes are often highly chaotic: there are many parties, many procedures, many issues. A process design may offer a certain amount of transparency. It allows the parties involved to inquire at any time where they stand in the decision-making process, what the nature of a decision is, and so on.

### *2.7.5 De-politicizing Decision Making*

Change processes tend to provoke much resistance. Excessive substantive steering at the start of a change process can stimulate resistance. A process approach to change can reduce this resistance, since it does not specify the substance of the change, only the process towards a possible change. (As one of the arguments, Giddens [14] notes that this creates trust among the parties).

Let us look at a simple example: the distribution of offices and equipment within a company. One could note that new offices and new equipment are always allocated to the people who are most proficient at playing the decision-making game: they are the most aggressive and the best informed, and they possess the power instruments to serve their own interests. In other words, the decision making is highly politicized. Those who present substantive arguments every time new space or equipment becomes available, and who challenge the established interests, run the risk of evoking serious conflict.

An alternative is a process intervention. This may imply, for instance, that a process is started to define criteria for distributing space and equipment. This should be done at a time when new space or equipment are not available. The members of the organization are involved in this process, and it should lead to consensus regarding these criteria. There are two ways in which such a process is de-politicizing. Firstly, *reframing* the agenda—'which are reasonable criteria?' rather than 'who gets what?'—leads to a different kind of discussion, which is less driven by the limited interests of the players in the organization. After all, the players suddenly find themselves participating in a discussion, which is about criteria and argumentation rather than about managerial aptitude. Secondly, once the criteria have been formulated, the substantive decision making—who gets which new space and equipment—will become easier, and less dominated by established power relationships [22].

### *2.7.6 Support*

Decision making involves a large number of parties, which often have *obstructive power*: they have the ability to stall the decision making, sometimes for prolonged periods of time. These parties will only provide their support if they are involved in the process of problem definition and solution.

## 2.8 The Result of a Process: Consensus, Commitment or Tolerance

Of course the result of a process partly depends on its goals. The result may for instance be a number of decisions, a number of actions, a physical product, a strategic plan, and so on.

There may be *consensus* about the results of the process: the parties involved fully agree with each other. The process has been fair and the substantive result has everyone's approval.

In most cases, however, there is such a considerable conflict of interests that consensus cannot reasonably be expected. In that case, a process can still lead to parties' *commitment* to a particular result.

A party that makes a commitment declares that, although it commits itself to the result, it does not do so on substantive grounds. A party may declare such a commitment because, for example, it has learned during the process that it has no alternative. Non-commitment would be more harmful than commitment.

Commitment implies that a party is willing to contribute to the implementation of decisions. Alternatively, a party may state that it will not declare any commitment (and thus will not contribute to the implementation of a decision), but that it will *tolerate* the results. This implies that it will not obstruct or hamper the implementation of a decision. Why? For instance because a party does not want to get into trouble with other parties, or it may be compensated elsewhere.

## 2.9 The Risks of a Process Approach

Which are the risks of a process approach? (Part of the following is derived from de Bruijn [7]).

*Risk 1: Explaining Rather than Discussing* This implies that process management is regarded as an instrument to communicate an already taken substantive decision as effectively as possible. In this line of reasoning, process management is not a means to confer with other parties about decisions that have to be taken, but it merely serves to explain future decisions properly. In this case there is already an adequate and deliberate decision, but it still evokes resistance. In order to overcome this resistance it is helpful to use a number of process management techniques.

In organizations, a concrete manifestation of this is the fact that process management tends to be the responsibility of the department of communication. This is because process management has been narrowed down to adequate communication: providing a proper *ex post* explanation of the sensible decisions an initiator has taken.

*Risk 2: A Project-Type Template for the Process* A second risk is that processes are shaped by means of a project-type template. In such cases, a process is designed with such tight goals, preconditions, budget and planning that there is only limited room for consultation and negotiation. In addition, the sequence followed is that of a project: there is a phase of problem exploration, the next phase serves to determine the goals, then there is a phase of information collection, then a decision is taken, which is implemented and evaluated during the next phase. The process proceeds according to the unshakable logic of project management, albeit with involvement of parties in every phase—within predetermined boundary conditions. This forces the process into a project-like template: once parties have joined the process, they can move in only one direction—towards the next phase—and have limited degrees of freedom.

Much of the disappointment about interactive decision-making processes can be traced back to the project-type implementation of these processes. First, a number of non-negotiable boundary conditions are defined, then the process approach is shaped by means of a project-type template. This leaves hardly any room for the actors involved, despite the expectations raised by the use of the correct process language.

*Risk 3: Process Management Results in Sluggish Decision Making* A major risk of process management is that it leads to slow and sluggish decision making. If many parties are involved in the decision making, there are many possibilities for these parties to block this decision making. If a manager indicates that he highly values support and organizes a process for that reason, this may even be an incentive for the parties to delay the process. If all parties are involved in the decision making and no decision is taken before there is support, parties will benefit from blocking the decision making. After all, as long as there is no support, there is no decision. A process approach thus gives rise to typical strategic behaviour.

Let us take another look at the decision making regarding the problems of Rotterdam. On the one hand, the supporters of a Maasvlakte area will join in this process. They will send representatives with a mandate. Opponents have much less interest in this process. There is a significant chance that they will send representatives with a limited or no mandate, who have no interest in the advancement of the decision making. These opponents will benefit from obstructing or hampering the process continuously, which leads to the process resulting in the opposite of what has been envisioned: there is no support, and thus no advancement of the decision making. Instead, it is delayed or even blocked.

*Risk 4: Process Management Leads to Impoverished Decision Making* It is often assumed that process management leads to substantively enriched decision making. An important risk, however, is that the exact opposite is true.

Firstly, if many parties are to be involved in the decision making, the result should do justice to the interests of these parties. Chances are that the quality of these results will be below average: it is a tasteless compromise.

Secondly, there is a risk of bargaining. Losers are compensated more or less randomly, for instance when they block the decision making at a certain point in time. Consequently, the eventual decision consists of a package of issues that are not substantively connected.

There is a difference between bargaining and negotiating. Bargaining is like horse trading: issues are grouped randomly in order to reach a deal. Negotiating also implies that certain issues are grouped, but in this case the parties' criterion will be that this grouping must result in synergy and enrichment, and that the negotiated result must be satisfactory and enriched.

Thirdly, enrichment takes place when ideas are measured against each other. A manager who is fascinated by the process approach runs the risk of limiting his own role to organizing and managing the process. If it is true that enrichment takes place when ideas are measured against each other, such an attitude towards process management will not suffice. Process-like management without substance fails to inspire and to provoke opposition, and therefore it will not result in enrichment.

## 2.10 Further Observations About Processes

So far, we have made some observations about the positioning of, arguments for and risks of a process-like approach to change. Let us now take a closer look at the kind of processes that we are referring to. After all, everything in life may be a process, and therefore a further description is useful. A process implies movement, change. Therefore we will start with a reflection on change. We will then clarify the concept of change by addressing:

- the relationship between change and negotiations about change;
- the relationship between decisions about change and their implementation; and
- the relationship between process and procedure.

### *2.10.1 Change*

What is change? The answer depends on the perspective taken. Let us look at a number of possible and relevant perspectives.

#### 2.10.1.1 Time Perspective

One person may interpret a series of events and decisions as change, while another may only see a static situation. This difference between perception and assigned meaning may relate to the time perspective. From a short-term perspective, changes are often hardly noticeable. A longer-term perspective will reveal patterns, and these are easier to identify as change.

> The European Union has become what it is today through a process that has taken over fifty years. Those who compare today's EU to the fragmented Europe of fifty years ago can only conclude that much has changed. Those who look at one year after the next, however, may easily conclude that there is little or no change. Managers and civil servants who actively participate in such processes will tend to take a short-term perspective. They will perceive the situation as inert and difficult to change. Historians who take a longer-term perspective, however, may reach another conclusion. In retrospect, they could characterize the second half of the 20th century as a highly dynamic period, with many changes in a relatively short period of time.

#### 2.10.1.2 Scope

With regard to the question whether a particular EU member state has changed significantly due to the development of the EU, various perspectives are possible. One may either focus on one single aspect or on several different ones. Potentially relevant aspects include social security, employment, economy, and so on. A one-aspect analysis is more likely to reach a radical conclusion, for instance that little—or much—has changed. A multi-aspect analysis will yield a more moderate result. After all, a broader spectrum of changes will tend towards an average.

#### 2.10.1.3 Scale

Scale is another aspect that determines whether or not change is apparent. A lower level of scale may reveal changes that differ from those at a higher level. At the level of one company, for instance, there may be a significant change, while at the higher level—the economic branch as a whole—the change is much less obvious.

#### 2.10.1.4 Population Versus 'Delta'

Changes in a system can easily be either trivialized or overrated. Trivialization may happen when one-first describes the entire system, and then characterizes the change in the context of the system as a whole. This will almost always demonstrate that the population as a whole has hardly changed, even though some individuals may show some minor changes. Another way of describing the change is by focusing on the individuals that have changed, rather than on the entire population. This may result in a change being overrated. For instance, when describing changes in our energy use in sustainability terms, one can describe the entire energy system and then conclude that only a small fraction of our energy sources are sustainable. In this case the focus is on the population. Another perspective results from zooming in on particular changes in the system, for instance the increasing share of wind energy in some submarkets. The changes in these submarkets can be substantial. In this case, the focus is on the absolute change—or 'Delta'.

#### 2.10.1.5 Perspective

Many change processes are goal-driven. The goal may for instance be an increased efficiency, a larger market share or a more sustainable production. Suppose that there has been an actual intervention to implement these envisioned changed. Budget streams have been altered, organizations have been restructured or rules have been changed. After certain period of time, there will be an *ex post* evaluation to determine whether the envisioned change has really taken place. Regardless of the way this question will be answered, the changes will be larger if a broader perspective is taken. If not just the envisioned change is taken into account, but all additional changes as well, the actual change will turn out to be more substantial.

Table 2.3 shows an often-used typology of change processes. When only taking into account the changes in the top left-hand cell, one may conclude that the changes are moderate. When looking at all four cells, one will be able to observe quite a few more changes.

**Table 2.3** Perspectives to change

| | Desired changes | Additional changes |
|---|---|---|
| Foreseen changes | | |
| Unforeseen changes | | |

## *2.10.2 Negotiation and Change*

This book is about 'negotiated change', in other words, changes that are the result of negotiations. Complex change processes that are being studied over a longer period of time tend to be a combination of 'real world changes' and 'paper changes'. Real world changes are changes at the level of output in terms of rules, laws, organizations and of course societal outcomes. In addition, there are changes that are agreed upon at the negotiation table. Parties may agree, for instance, to change their behaviour. This agreement may be formalized in documents (such as minutes, treaties, agreements and so on). These also qualify as output. In addition, and partly because of this, changes are implemented in the 'real world'.

In this kind of processes, 'negotiation' should not be taken too literally. Of course, parties negotiate—they are sitting around a table, they follow an agenda and there are journalists who are waiting for an outcome. But in addition, there are more implicit negotiations that are at the basis of real world change. These implicit negotiations are followed by tacit agreements, and may proceed in several different ways. Parties may, for instance, engage in unilateral communication. In that case a party makes statements, gives interviews, or publishes studies that express its views. In turn, the other party reacts by giving interviews, producing reports and so on. This may eventually lead to tacit agreements. Parties may anticipate each other's preferences, upon which other parties take action to address the first party's interests, and so on. We regard these movements as an important aspect of processes. It should be emphasized that these processes sometimes remain unanalyzed when the perspective taken is purely that of negotiation.

Regarding the tension between negotiations and change, it is also important to take into account the differences between *outputs* and *outcomes*. Outputs may be the direct result of a negotiation. An output may be, for instance, the amended text of a law that was agreed upon during a negotiation. A cynic may say: 'So what, a law alone does not change anything.' But those who make a more detailed analysis will understand that this amended law makes 'real world changes'—the *outcomes*—much more likely, and sometimes even inevitable, although it may take some time before the *outcomes* are actually implemented.

Special attention should also be paid to the process of assigning a joint meaning to certain concepts as a consequence of negotiation processes. During their negotiation, parties arrive at statements that define a societal situation or development. For instance, they issue a joint statement that declares that 'the current situation certainly carries certain risks that may lead to major problems'. Such a statement can hardly be called an output, let alone an outcome. Yet the meaning of such a statement may have important implications for the future actions of these parties.

### 2.10.3 Deciding and Implementing

The term 'change process' may have two different meanings:

The first is the decision-making process with regard to the change. These processes tend to take place not only within formal decision-making fora such as parliaments or Boards of Directors, but also—often: particularly—in informal, ad hoc-like processes in task forces, project groups or—even more diffuse—in spontaneously arising, network-like configurations of actors that tend to change rapidly in terms of shape and intensity. Activities that manifest themselves in this perspective include consultations, negotiation, research, reporting, media contacts, and the continuous hassle and switch between speeding up and slowing down that seems to characterize the work of many managers and their staff.

The second meaning is the change itself, for instance institutional changes, reorganizations, physical changes such as city expansions and infrastructure transformations, and so on. These can be regarded as the implementation of the decision that has been taken. The implementation itself is often a process as well.

The EU has taken over fifty years to become what it is today. Politicians such as Schuman, Monnet, Churchill and Adenauer, who took the first few steps towards closer cooperation, could not have foreseen that in 2008, 27 European countries would cooperate intensively, supported by a European Parliament, a European Commission, its own official set of rules and policies, and even—to a limited extent—its own European currency. They could not have predicted how this process would move forward: sometimes fast, sometimes extremely slowly. Sometimes with success, in an almost euphoric atmosphere, and sometimes in an atmosphere of crisis.

When it comes to the first meaning of 'process', the focus is on the decision making about the EU—in other words, on the many commissions that were established, the many 'crucial' conferences that were held, the plans that were launched, and so on. In this perspective, it is relevant to wonder which of these managerial activities have resulted in which outcomes.

With regard to the second meaning, the focus is on the actual institutional changes that have taken place. In this perspective, the 'real' change is key—for instance the shift in competencies, the changes in financial flows, new European guidelines or the new European currency. These actual changes are what really matters, and it is possible to analyze how they were accomplished.

The decision making about the establishment of the EU as well as the institutional changes itself may be regarded as a process.

A second example is the process that has resulted in the expansion of infrastructural networks. During the last few decades of the 19th century and the first few decades of the 20th century, initial ideas were launched about major infrastructural works, such as sewage, drinking water and railway systems. From the perspective of the first meaning, attention focuses on the plans for these networks, on the fights to secure financing for their implementation, on the efforts by supporters and opponents of these infrastructures, and so on. The next question would be which of these actions have actually led to concrete implementation.

The second meaning addresses the actual growth of the networks in the course of several decades. The actual development is key, and analysis focuses on how this actual development came about and how it can be explained on the basis of the relevant decision-making processes.

Why does this difference matter? Of course the two meanings tend to be correlated. Decision making about change (meaning 1) results in a change (meaning 2). However, in many cases reality is different. Many managerial activities (meaning 1) do not actually result in 'real' changes (meaning 2) and vice versa: 'real' changes (meaning 2) cannot necessarily be traced back to decision-making processes (meaning 1). The word 'process' in process management refers to both of these meanings. Qualifications like 'a good process', an 'incremental process' or a 'politicized process' may refer to the managerial process as well as to the actual change process.

Why do we emphasize this? Because many sources still refer to decision making or policy making on the one side, and implementation on the other side. The difference between the two, however, is very problematic in network-like environments with a large number of actors.

In their seminal essay called *Implementation*, Pressman and Wildavsky [24] show that any decision will undergo quite some revisions during its implementation. They show that many actors are involved in the decision's implementation, and in the end hardly any aspect of the decision remains unchanged if each of these actors is allowed to slightly amend the decision. As a result, the eventual outcome hardly relates to the taken decision, while this cannot be attributed to one single actor. They conclude that decisions (or: policies) are shaped during their implementation, which is why there should be no radical distinction between decision making or policy making on the one side and implementation on the other side. This is a first effort to put into perspective the importance of The Decision as it was taken at a given moment. Apparently it is not just the decision that matters; apparently there is not one single moment at which the policy is made. Rather, the policy is shaped at several moments, over a longer period of time [1, p. 252]. If this is true, processes deserve more attention. It is in processes between actors that policy is shaped and decisions are made.

Pressman and Wildavsky do take an actual decision as a starting point. The decision may well change during its implementation, but it remains more or less a definitive decision. Teisman takes this notion one step further [28]. In his research into the decision making about infrastructures, he concludes that there is no empirical evidence for 'the decision' about the realization of the infrastructure. There is either no decision, or there are many decisions. Looking back from the point in time when the infrastructure has been implemented (meaning 2), it will be hard to identify the definitive decision. But what can be identified then? There may have been decisions about finances—funds have been made available—but this does not imply that the infrastructure will actually be realized. The sponsor will therefore not consider himself as the actor who allowed the infrastructure to be realized.

Of course there have also been decisions about the spatial implementation of the infrastructure: the regional plan and the land use plan will have undergone certain changes so as to allow the spatial realization of the infrastructure. Neither does this decision constitute the actual decision that the infrastructure would be realized. It is as if this kind of decision eliminates obstacles to the realization of the infrastructure, but no more than that. None of the actors feel as if they made the all-inclusive and definitive decision.

Despite the fact that *the* decision cannot be identified, the infrastructure has been realized. The infrastructure has, so to speak, *emerged* during the process. The decision itself is no more than a theoretical construction. This is the reason why this book does not focus on decisions per se, but rather on the processes in which decisions are made and that result in actual changes.

### 2.10.4 Process and Procedure

Processes and procedures may be closely related, but they are not the same thing. Processes can develop spontaneously and they have no fixed shape—the players determine the rules of the game. Procedures, on the other hand, have been laid out, for instance in laws or regulations. Procedures are characterized by pre-described phases, which may have a minimum or maximum duration, and by a well-defined order of activities. Procedures have a legal framework; deviation from procedures may cause major damage.

Procedures are usually a part of the process. Procedures appear later on in the process, once the ideas in the process are further defined.

Decision making about infrastructures is a process. For years, there may be rather indeterminate talk about a certain infrastructural connection. Once the impression of the infrastructure is consolidated and parties' commitment to the infrastructure is further determined, formal procedures are started. The procedure to raise funds, the procedure for an environmental impact assessment, a change of the regional plan, and so on.

Process and procedure may well go hand in hand. In that case, the procedure is ongoing, while the intermanagerial process proceeds in parallel. Managers continue to negotiate about the infrastructure: in which variety will it be implemented, who will pay for it, how will it be utilized, and so on.

The transition from process to procedure is not always smooth. Certain parties may hold a strong position in a procedure that has to be followed, while their position in the process may only have been minor. Such parties include those that did not participate in the process, but should be consulted, or even have approving authority, according to the procedure. This way, parties may gain influence on the process, which they did not have at first. It may be a question of good process management to anticipate the inevitable procedures already during the process phase. The strategy may also be to generate such massive support and momentum during the process that the procedure becomes a mere walkover.

The way in which process and procedure are coupled may be part of the process architecture, and is a key focus area for the process manager.

## 2.11 Process Management and Related Approaches

This section will address the positioning of process management vis-à-vis related approaches to decision making. How do our definition of and attitude towards process design and process management relate to procedural rationality, consensus building and interactive decision making?

Our description of these approaches does not aim to be exhaustive; we will only use the comparison with the related approaches to position process design and process management more accurately.

### *2.11.1 Procedural Rationality*

In his seminal article called 'From Substantive to Procedural Rationality', Simon [26] introduces the distinction between substantive rationality and procedural rationality. Simon identifies behaviour as substantively rational if it contributes to the realization of given objectives; behaviour is procedurally rational if the process that results in a decision is 'correct'. Simon's definitions of 'correct procedures' include adequate thinking processes and the use of the correct algorithms.

The similarity between process management and procedural rationality is obvious. Both approaches accept that in many cases it is impossible to design the best substantive solution for a substantive problem. This is why in both approaches the choice is made not to continue searching exclusively for substantive solutions to problems, but also to take into account the correct process. This should then result in solutions that are substantively satisfactory. Both approaches are based on the assumption that a well-designed process results in a substantively good solution. There are, however, also important differences between the process approach and the concept of procedural rationality.

#### 2.11.1.1 Process as a Rational Design or as a Result of Negotiation

Firstly, according to Simon, the correct process is again a matter of rational design. In his view, there is an ideal procedure to approach a particular problem. He therefore refers to *procedural rationality*. Finding and designing the correct procedure is therefore, in his opinion, primarily a scientific exercise. Once the best procedure has been designed, it can be prescribed. Where appropriate, problems should be solved by means of that process.

In the process approach, the process design is the result of negotiation. The relevant parties develop a certain level of commitment to a process design because they feel the process is fair and offers them sufficient opportunities to realize their own interests. Every actor makes an individual decision as to whether this is the case for a given draft process design. The fact that every actor arrives at an individual judgment about the quality of the process does not seem to correspond with the notion of a process that is objectively the best process thanks to a scientific design. A process is good because it is acceptable and trust-inspiring for the main stakeholders. In short, process management does not imply that a process is designed only once, or that it is universally and perpetually applicable when addressing a particular kind of problem.

Secondly, the parties that are supposed to arrive at a solution in a network, will *learn*. When a certain party attempts to influence another party by applying a certain management tool, the other party will learn how to deal with this instrument and also how to evade its steering effect. In the context of a network, the effectiveness curve of an instrument will therefore look as follows (Fig. 2.1):

The more one single process design is used, the more the relevant parties will learn how to evade its steering effect. They will learn, for instance, to identify the weaknesses of a process design, and use those to advance their own interests. This may diminish the value of the process for other actors. As a result, the effectiveness and usefulness of the same process design will gradually decline. This Law of Diminishing Effectiveness implies that there is no such thing as a 'best' process that will always remain the best. Every process design will cease to be effective at some stage.

Diminishing effectiveness also applies to various strategies that parties may use in the process. The following three examples, derived from Cialdini [4], will illustrate this. Firstly, *charm and disarm*: those who are positive and flattering towards others usually increase the likelihood that these other people will join an agreement. Involve *old-timers*: authoritative senior players who share their opinion, making it difficult for other parties to ignore that opinion. Provide *exclusive information*: those who feel they are ahead of others in terms of the information that they have, will use this information and thus be more likely to join an agreement. Regardless of the strategy used, those who have dealt with the situation a few times will learn what the strategies are, and also how to invalidate them. How? For instance by describing these strategies in explicit terms, allowing others to recognize and anticipate them.

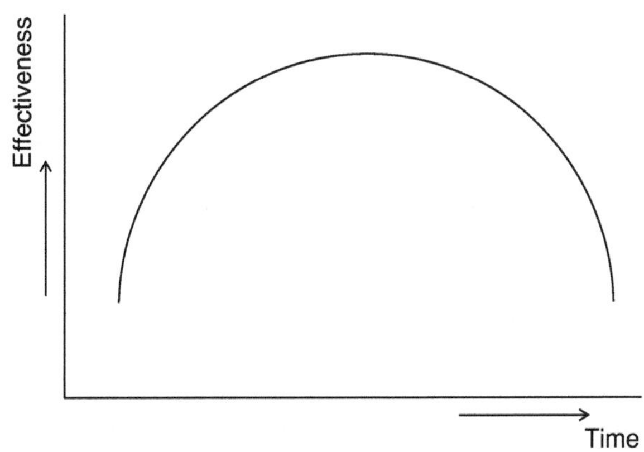


**Fig. 2.1** The effectiveness of instruments in a network

### *2.11.2 Consensus Building and Mediating*

Process management is of course closely related to consensus building, which may be defined as 'a process of seeking unanimous agreement. It involves a good-faith effort to meet the interests of all stakeholders' [27, p. 7]. One of the key techniques of consensus building is to cast parties' positions in more abstract terms, and focus attention at their underlying interests. The 'manager' plays a facilitating and mediating role. *Facilitation* may for instance include methods to stimulate groups to communicate effectively face-to-face. *Mediation* aims to bridge the gap between parties with highly contrasting views. Mediation may take a considerable amount of time. It aims to prevent a 'zero sum game', and to steer the process towards new solutions that have an added value for all parties.

Such consensus-building techniques may be an integral part of process management. However, there are some important differences.

#### 2.11.2.1 Process Management is Embedded Both Institutionally and Administratively

Consensus building is presented as universally applicable. The method comprises a large number of steps that have to be taken in order to reach consensus. In many—though not all—cases, the focus is on the interaction between people.

Process management, on the other hand, is embedded in existing administrative processes. In these processes one may identify all sorts of phenomena, which are a part of the managerial game.

- When it suits them, parties will practice power play. They use their dominant position to push their opinion through. Consensus may also be forced (as will be demonstrated in Chap. 7).
- Parties are aware of the fact that they will meet again during future consultations. They may swap the results of the negotiation in the current process for future results regarding another issue.
- Parties play strategic games: they give misrepresentations, play a waiting game, want to keep all options open above anything else, have no interest in consensus, and so on.

Since process management is embedded in an administrative process, it should be adapted to such types of behaviour. This largely implies that process management allows parties to show their natural behaviour. A party that aims to work with a hidden agenda should be allowed to do so; there is no such thing as a ban on hidden agendas. Nor does it make any sense to establish the rule that parties have to be open towards each other during the process or study each other's interests. It is wiser to allow parties sufficient room to show their natural behaviour.

In this case, too, the fact that process management is embedded in administrative/managerial processes implies that there are no universally applicable process designs. Process design is shaped according to the specific administrative/managerial context.

### 2.11.3 Interactive Decision Making

In interactive decision making, the body that is competent to take a decision involves other actors, such as citizens, companies and interest groups, in the decision making [21]. Interactive decision making has gained much popularity in the context of the development of spatial plans, such as land use plans, regional plans and plans for the construction of infrastructural projects. Interactive decision making builds upon the tradition of participation in the development of (spatial) policy.

Interactive decision making has a number of features in common with process management. The first common feature is that the actor initiating the interactive decision making apparently wishes other actors to participate in the decision-making process. Something similar is true in process management. Here, too, there is an actor with an interest who wishes to arrive at a decision, and who wishes to enter into an exchange of ideas with other actors. The second common feature is that both interactive decision making and process management are highly contingent. Interactive decision making, too, is shaped depending on specific questions that arise, and depending on the nature of the actors whose interests are at stake.

#### 2.11.3.1 Process Management is Administratively Oriented

However, there are also a number of differences between process management and interactive decision making. In the interactive decision-making practice, it is usually a public authority that develops a plan in cooperation with relevant citizens, companies and societal organizations. Process management, on the other hand, is oriented towards actors who cooperate in administrative or managerial processes. In other words, process management has a strong administrative or managerial focus, while interactive decision making has a societal focus.

Interactive decision making tends to start with a government that has to take a decision or make a plan. This is why it is this government's task to design the interactive process: the government determines the rules for the process, defines the substance of the process, decides when the interactive process will start and during what period the interactive decision making will take place. In process management, on the other hand, establishing the process is an activity in which several parties are involved. The latter is due to the fact that interactive processes hardly ever concern decisions that have to be taken. They usually concern plans, views or policies. If they concern decision making at all, the relation between the processes and the prospective decisions is very indirect. There is a chance that interactive decision making thus becomes an activity with no strings attached.

**Table 2.4** Process management compared to related approaches

| Related methods | Similarity to process management: both … | Difference with process management: process management … |
|---|---|---|
| Procedural rationality | … postulate that only a process can be designed, not its substance | … involves a process resulting from negotiation;<br>… does not involve 'preservable' processes |
| Consensus building | … concentrate on interests and on avoiding 'zero sum games' | … is administratively embedded and tolerates administratively strategic behaviour |
| Interactive decision making | … are open as well as contingent | … is administratively oriented and less non-committal |

Managerially intelligent actors who are aware of this will also know that participation in interactive 'decision making' has limited significance. They will either participate in a disorderly manner, or not participate at all. Interactive decision-making processes tend to have a moral connotation as well: from a normative point of view, it is desirable to involve parties in the 'decision making'. This is much less the case for process management; instead, it is an approach that may be useful when decisions have to be taken in a network and when parties know that they depend on each other. Parties will only join a process if it has something to offer them; consequently, the process should indeed concern decisions that have to be taken.

Since process management is about decisions that have to be taken by administrative parties, much attention is being paid to the natural strategic behaviour of these parties, as mentioned before. Interactive decision making, however, takes place against a different background: parties are supposed to adopt an open attitude and be receptive to each other's interests (Table 2.4).

## References

1. Barzelay M (2004) The process dynamics of public management policymaking. Int Public Manag J 6(3):251–281
2. Carpenter SL, Kennedy WJD (2001) Managing public disputes. practical guide for government, business, and citizens groups. Wiley, New York
3. Chisholm D (1989) Coordination without hierarchy. University of California Press, Berkeley
4. Cialdini RB (2001) Harnessing the science of persuasion. Harv Bus Rev Point Summer 79(9):72–79
5. Cohen MD, March JG, Olsen JP (1972) A garbage can model of organizational choice. Adm Sci Q 17(1):1–25
6. Crozier N, Friedberg E (1977) Actors and systems. University of Chicago Press, Chicago
7. De Bruijn JA (1999) Van sturing tot proces. In: in 't Veld RJ (ed) Sturingswaan & ontnuchtering. Lemma, Utrecht, pp 52–68
8. De Bruijn JA (2000) Processen van verandering. Lemma, Utrecht

9. De Bruijn JA, ten Heuvelhof EF, R.J. in ‘t Veld (1998) Procesmanagement: Besluitvorming over de milieu- en economische aspecten van verpakkingen voor consumentenprodukten, Delft
10. Douglas M, Wildavsky A (1982) Risk and culture. University of California Press, Berkeley
11. Dunn WN (1981) Public policy analysis: an introduction. Pearson Prentice Hall, Upper Saddle River
12. Fisher R, Ury W (1981) Getting to yes: negotiating agreement without giving in. Houghton Mifflin, Boston
13. Gerrits LM (2008) The gentle art of coevolution. Erasmus Universiteit, Rotterdam
14. Giddens A (1994) Beyond left and right: the future of radical politics. Stanford University Press, Stanford
15. Guinée J et al (2002) LCA, an operational guide to the ISO-standards. Kluwer, Dordrecht
16. Healey P (1998) Collaborative planning in a stakeholder society. Town Plann Rev 69(1):1–21
17. Hisschemöller M, Hoppe R, Dunn WN, Ravetz JR (eds) (2001) Knowledge, power and participation in environmental policy analysis: policy studies review annual. Transaction Publishers, New Jersey
18. Jordan AG, Schubert K (1992) A preliminary ordering of policy of network labors. Eur J Polit Res 21(1):7–27
19. Jordan AG (1990) Sub-governments policy communities and networks. J Theoret Polit 2(1):319–338
20. Kenis P, Schneider V (1991) Policy networks and policy analysis: scrutinizing a new analytical toolbox. In: Marin B, Mayntz R (eds) Policy networks, empirical evidence and theoretical considerations. Westview Press, Boulder, pp 26–59
21. Klijn EH, Koppenjan JFM (2000) Interactive decision making and representative democracy: institutional collisions and solutions. In: van Heffen O et al. (eds) Governance in modern society. Kluwer, Dordrecht, pp 109–134
22. Kolb DM, Williams J (2008) Breakthrough bargaining. Harv Bus Rev Point 79(2):39–47
23. Matthews WA, McKenzie RL (2006) Parallels and contrasts between the science of ozone depletion and climate change (unpublished)
24. Pressman JL, Wildavsky AB (1973) Implementation: how great expectations in Washington are dashed in Oakland. University of California Press, Berkeley
25. Sebenius JK (1991) Designing negotiations toward a new regime. the case of global warming. Int Scc 15(4):110–148
26. Simon HA (1976) From substantive to procedural rationality. In: Latsis S (ed) Method and appraisal in economics. Cambridge University Press, Cambridge, pp 129–148
27. Susskind L, McKearnan S, Thomas-Larner J (1999) The consensus building handbook. Sage, Thousand Oaks
28. Teisman GR (1992) Complexe besluitvorming: Een pluricentrisch perspectief op besluitvorming over ruimtelijke investeringen. Elsevier, Den Haag
29. Teisman GR (1997) Sturen via creatieve concurrentie. Katholieke Universiteit Nijmegen, Nijmegen
30. Van den Donk W (1998) De arena in schema. Koninklijke Vermande, Tilburg
31. Vaughan D (1996) The challenger launch decision: risky technology, culture, and deviance at NASA. University of Chicago Press, Chicago
32. Watkins M (2003) Strategic simplification: toward a theory of modular design in negotiation. Int Negotiat 8(1):149–167
33. Watkins M, Lundberg K (1998) Getting to the table in Oslo: driving forces and channel factors. Negotiat J 14(2):115–135
34. Watkins M, Rosegrant S (2001) Breakthrough international negotiation: how great negotiators transformed the world’s toughest post-cold war conflicts. Jossey-Bass, San Francisco
35. Willke H (1993) Systemtheorie: Eine Einfuehrung in die Grundprobleme der Theorie Sozialer Systeme. Lucius & Lucius, Stuttgart

# Part II
# Process Architecture

# Chapter 3
# Designing a Process

## 3.1 Introduction

This chapter addresses the requirements of a good process. Some of these requirements can be met through the right process design. This is where we enter the domain of negotiation architecture [22, 29]. The structure of the chapter is as follows. Section 3.2 will introduce the four main requirements of a process, or process agreements (we will refer to these as the core elements of a process design). A good process is:

- an open process,
- in which parties are offered security through protection of their *core values*,
- which offers sufficient *incentives for progress and momentum*, and
- which offers sufficient guarantees for the *substantive quality* of the results.

We will translate each of these core elements into several more detailed design principles (Sects. 3.3–3.6).

## 3.2 The Four Core Elements of a Process Design

Our classification of the four core elements of a process design has the following rationale.

1. *Openness*. Process management means that an initiator does not take unilateral decisions, but adopts an open attitude. Other parties are offered an opportunity to participate in steering the decision making, and therefore also to highlight the issues they are interested in and that they feel should be placed on the agenda. Openness therefore concerns both the choice of participants and the decision-making agenda.
2. *Protection of core values*. Openness is not always appealing to parties invited to participate in a process. Every party will have its own interests, and runs the

H. de Bruijn et al., *Process Management*, DOI: 10.1007/978-3-642-13941-3_3,

risk that these interests are not sufficiently addressed. At the end of the process one or several parties may therefore not be satisfied with the result, while it is difficult for them to withdraw from the process at that stage. This is why there is a second category of design principles that results from the idea that the parties that commit themselves to a process—thereby taking a certain risk, perhaps even sticking their necks out—must be offered sufficient protection. How? They must be certain that their core values will not be harmed, regardless of the outcome of the process.

3. *Progress*. The first two core elements offer insufficient guarantee that a decision-making process will be good. If open decision making is opted for (core element 1), and parties' core values are protected (core element 2), chances are that even if there is discussion and negotiation, still no decision is made. Perhaps the outcome will include nothing but sluggish processes that will never produce a clear result. A third category of design principles addresses the need for the process to show sufficient *momentum and progress*.
4. *Substance*. Parties participating in an open process (core element 1) should be given sufficient protection of their position (core element 2), while there should also be sufficient guarantees that progress will be made in the decision-making process (core element 3). As a fourth requirement, this progress should meet certain substantive quality standards. After all, there may be strongly conflicting interests that force parties to make decisions that are substantively poor and perhaps even incorrect. Therefore it is crucial that the process has a sufficient number of substantive elements (Fig. 3.1).

A process design will always have to do justice to the four core elements, and will therefore always be a trade-off between those four core elements. A process without openness will be regarded as a concealed kind of project management and command and control. A process that does not protect parties' core values will be perceived as very unappealing and unsafe. Chances are that

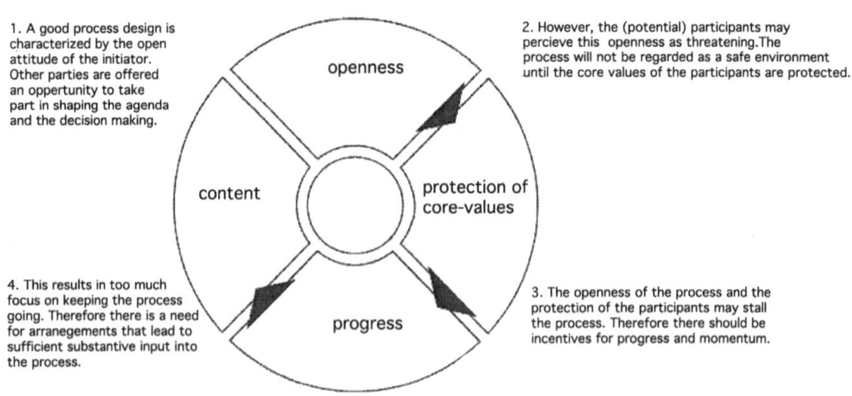


**Fig. 3.1** The four core elements of a process design

**Table 3.1** Design principles

| |
|---|
| Openness |
| 1. All relevant parties are involved in the decision-making process |
| 2. Substantive choices are transformed into process-type agreements |
| 3. Both process and process management are transparent |
| Protection of core values |
| 4. The core values of parties are protected |
| 5. Parties commit to the process rather than to the result |
| 6. Parties may postpone their commitments |
| 7. The process has exit rules |
| Progress |
| 8. Stimulate 'early participation' |
| 9. The process carries a prospect of gain |
| 10. There are quick wins |
| 11. The process is heavily staffed |
| 12. Conflicts are addressed in the periphery of the process |
| 13. Tolerance towards ambiguity |
| 14. Command and control are used to maintain momentum |
| Substance |
| 15. Substantive insights are used for facilitation. The roles of experts and stakeholders are both bundled and unbundled |
| 16. The process proceeds from substantive variety to selection |

parties, driven by mistrust, will keep delaying the process, or that they will not even join the process in the first place. If there are no arrangements to provide the process with progress and momentum, the process will become sluggish and may lose its authority. If there are no procedures to create substance and quality, a process may produce poor results that are vulnerable to outside criticism.

Sections 3.3–3.6 will describe the design principles. For the sake of clarity, Table 3.1 presents an overview of these.

## 3.3 Design Principles Leading to Open Decision Making[1]

### *3.3.1 Party Involvement: all Relevant Parties should be Involved*

The first design principle is that all relevant parties should be involved in the decision making. A key question is of course who the relevant parties are. Let us first make a simple classification of the different types of parties:

- Parties that have productive power and parties that have obstructive power. Parties with productive power have the means to actually (help to) implement an

[1] Refs. [1, 2, 8, 12, 13, 17].

initiative. These means may include finances, competencies, relations, physical resources, or expertise. Obstructive power means that parties are only in a position to obstruct an initiative.
- Parties that support an initiative, parties that oppose it, and parties that have not yet taken any particular position.
- Large parties and small parties.

Which of these parties are actually relevant? There is a tendency to invite mostly large parties that have productive power and that support an initiative. The risk will be evident: such an invitation may be an incentive for the other parties to join forces in their resistance against this initiative. Parties with obstructive power will also have to be involved in the decision making. Involving these parties in a process may prevent them from utilizing their obstructive power and thus from hampering the decision making. It may also be sensible to invite small parties. They often have the sympathy of other parties. Exclusion of small parties may stimulate others to support these small parties.

Does this mean that everyone should be invited? This will often be impossible, if only because it can stall the decision making—the process can result in utter indecision. Three considerations are important in this regard.

Firstly, the total of parties being invited should be an accurate representation of the parties that have an interest in the decision making. There are, for instance, many environmental organizations: activist, moderate, one-issue organizations, and so on. It will be impossible to involve all of them in a process, but as long as an initiator invites one or two of these organizations, the environmental interest is represented. There may also be many small parties that have an interest in decision making. An initiator will have to invite at least some of these small parties in order for them to be represented. Moreover, this increases the legitimacy of the process: the environmental interest and the small-party interest are represented, respectively. Chapter 5 will address this issue in more detail.

Secondly, there are different options for involving different parties in a process. A process may, for instance, have several different phases; one may choose to invite fewer parties in the initial phase, and more parties later on. It is also possible to identify different roles: some parties are involved in the decision making, while others provide obligatory advice or participate in the process as experts. This, too, will be discussed in Chap. 5.

Finally, when assessing whether or not to involve a party in the decision making, one should take into account other considerations besides just those relating to power positions. There may be *important moral arguments to involve certain parties in the decision making.* In almost every process there are parties that have no significant power position, but that may be affected by a potential decision. Those are the weaker parties that deserve protection.

### 3.3.2 *Process Agreements as a Means to Make Substantive Choices*

A second design principle is minimization of the number of substantive choices made prior to the process. Although the substantive issues may be assessed, this merely leads to an indication of how the decision making *process* will proceed. In other words, there is a transformation from substantive choices to process agreements.

In Chap. 2 we presented the example of the City of Rotterdam that intends to build a Second Maasvlakte area. When parties are invited to join a process that is to result in the construction of the Maasvlakte (i.e. a substantive choice), this will hardly be appealing. Instead, a number of process agreements may be made, such as:

- parties will commission research into the use and necessity of land reclamation;
- parties will commission research into a potential strengthening of the ecological structure of the region;
- parties use this research to determine their position regarding land reclamation and ecological strengthening of the region.

There are three advantages to this set of procedural agreements—which are only presented as an example here. Firstly, it offers enough room to all of the parties. The process will not be like a funnel trap that eventually forces parties to agree with a Maasvlakte. Secondly, when parties commit themselves to these agreements, they are sitting around the table—which is always better than a *free fight* between parties, which would have a high cost and little gain. Thirdly, the initiator also has a prospect of a result. If he plays the game well and knows how to deal with the interests of the other parties, there is a fair chance that he will realize at least some of his interests. And even though these agreements may seem little appealing to an initiator who is substantively driven and convinced of his case, they are often the BATNA: the Best Alternative To a Negotiated Agreement.

### 3.3.3 *Transparency of Both Process Design and Process Management*

Similarly, it is of key importance that the design of a decision-making process is transparent. Transparency implies that it is clear to parties how the process will take place, how their interests will be protected, which decision-making rules will apply, and—of course—who will be involved in the process. Transparency means that parties can check the integrity of the process, and whether or not it offers them enough opportunities. If a process lacks transparency and if parties are unaware of the process agreements, this will be a breeding ground for mutual distrust and, as a consequence, for conflict.

The role of the process manager should also be transparent. His role is that of an independent facilitator. Of course the process manager should focus on the process rather than on the substance: strong substantive views may be regarded as substantive prejudices by one or more parties, and thus as an infringement of the process manager's objectivity.

# 3.4 Design Principles that Protect Parties' Safety and Core Values[2]

## 3.4.1 Protecting Parties' Core Values

Openness means that the parties joining a process should be offered a chance to influence future decision making; it should be avoided that they feel as if they enter a funnel trap. This particularly applies to parties that are not easily persuaded to join a process. These parties should not get the impression that their participation implies that they will get trapped in a decision-making process, that the envisioned decision has in fact already been taken, and that their participation contributes to its justification. One way to prevent this is to offer the parties protection of their core values: they may then be sure that as far as these values are concerned, they will not be forced to adopt a certain behaviour or make certain choices against their will.

The following example will illustrate this. An environmental organization joining a decision-making process about new infrastructure cannot be expected to refrain from making press statements during the process. Core values touch upon the essence of the organization. A core value of an environmental organization is the fact that it has the opportunity to mobilize public opinion about environmental damage. An environmental organization will not be eager to participate in a process if it is forced to remain silent for a prolonged period of time, particularly when one keeps in mind that the outcome of any process is uncertain.

A core value should not be confused with a position that an organization adopts in a decision-making process. The core value transcends the level of the single process and the single point of view, and has a much more generic nature.

## 3.4.2 Commitment to the Process Rather than to the Result

As outlined in Sect. 3.3, parties cannot be expected to make substantive choices prior to the process. They only commit to a set of process agreements. This also touches upon the question whether or not parties should be asked to make a prior commitment to the result of a process. Even though this result is yet unknown, it seems reasonable to ask parties to commit to it—otherwise, after all, there is a chance that the initiator designs a time-consuming process only to conclude at the end of it that the parties remain divided. Moreover, parties influence the result. They participate in the process—which is why they may be reproached afterwards that they 'stood by and watched'.

[2] Refs. [15, 16, 20].

From the perspective of the other parties, however, the requirement of a prior commitment to the result is highly unreasonable. An important aspect of the protection of parties' core values is that they are *not* asked to commit to any process result beforehand. They can only be asked to commit to the process. Not committing to a result will create safety and space. It is this space that will allow parties to develop commitment to the result during the process. In that case, however, it will truly be commitment, rather than a forced commitment that will not be sustained anyway.

### 3.4.3 Commitments to Subdecisions may be Postponed

A decision-making process usually consists of a large number of subdecisions that will eventually lead to a final decision.

Let us take another look at the process regarding the expansion of the Rotterdam port area: the Maasvlakte. This process will be characterized by all kinds of subdecisions relating to, for instance, the area of land that might be reclaimed, various potential compensation projects, the potential construction of access roads, the potential elongation of a railway connection, the potential relocation of an airport, and so on. In this case, too, an initiator may want parties to commit to some of these subdecisions at a certain point in time. This, after all, will seem like an indication of progress. But the other parties may feel differently. Commitment to subdecisions may feed the notion of the process being a funnel trap. It may feed the notion that *points of no return* are being created. If this notion becomes well-established, the consequences will be obvious: there will be strong incentives for distrust and resistance.

An important aspect of safety and protection of core values is the fact that parties are not asked to commit to subdecisions during the process. Only at the end of the process will the parties be asked for their commitment to the final package of decisions. As the saying goes: 'Nothing is decided until everything is decided.'

### 3.4.4 There are Exit Rules

An important design principle is that a good process has exit rules: rules allowing parties to leave the process. The process agreements may stipulate, for instance, that the parties may evaluate after some time whether or not they wish to continue their participation in the process. For some parties, this lowers the threshold to join the process. Such an exit rule also greatly reduces the risk of participation for an individual party. After all, the party may leave the process even before the definitive decision making. This eliminates the funnel trap perception.

Of course the initiator as well as the process manager will have an interest in preventing parties from utilizing this exit option. In an ideal situation, participation in the process becomes sufficiently appealing that leaving is no longer an option. This will be further addressed in Chap. 6. The above-mentioned mechanism manifests itself again:

- the exit option creates safety and space,
- which will nourish cooperation and decision making.

If there is no exit option, it will be inevitable that some parties will not join the process, or that there will be serious conflicts during the process. It is the process—rather than a set of preconditions—that should do the work.

## 3.5 Design Principles that Guarantee Progress[3]

Processes may show little progress because parties make no prior commitment to the final result, nor to subresults, as has been pointed out in Sect. 3.4. This is inevitable, but not without risk. Parties within as well as outside of the process may get the impression that the process only leads to sluggishness and fails to produce any results. This may affect the legitimacy of the process, and thereby the chances of success. What can be done to stimulate progress and momentum?

### *3.5.1 Stimulate 'Early Participation'*

Change processes that require a wide participation in order to be successful are usually characterized by a slow start. This is true for instance with regard to the combat against global environmental problems, which many countries—if not all—should participate in. What appears to be necessary—i.e., all parties being in agreement and ready to start—is difficult to organize. There are many reasons for this. Firstly, there are always actors that disagree with the starting conditions, or for whom it is not convenient to start at the chosen moment. Secondly, countries' reluctant or evasive behaviour may well have a strategic nature. These countries will feel that as long as other actors adhere to the rules, they themselves can afford to reject the consensus. After all, the problem will also be solved if they are the only ones not to participate. This behaviour, even if it is shown by just a few actors, may paralyze the process. Parties that would have been willing to participate now see their position being affected by the reluctant behaviour of other parties, which may be a stimulus for them not to participate in the process either. Thirdly, parties that are willing to make agreements about behavioural changes early on during the process may well anticipate that they will be punished for this at a later stage, when other actors join the process. It is easy for parties that anticipate this to decide to 'temporarily' refrain from participating.

---

[3] Refs. [3–5, 26, 27].

Suppose some countries feel that a certain chemical substance is harmful to the environment and its use should be drastically reduced. In order to really solve this problem at a global scale, all or almost all of the countries should take part in this reduction, but this appears to be impossible to attain. Suppose these countries—let us call them *early participants*—are so motivated that they make agreements anyway: they all commit themselves to a 20% reduction over 5 years' time—despite the fact that the other countries' emissions will continue.

After 5 years, the countries meet again for evaluation and to make new agreements. Ideally, new countries will have joined in the discussions about a second round of reductions. If the discussions in the second round are successful, the parties will manage to agree to new reductions, for instance another 20% over the next 5 years. For these new parties, this 20% reduction will generally be easier to attain than for the parties that already participated in the first round. After all, the early participants have already achieved their 'easy reductions', and the next round of reductions will probably demand more sacrifices than the first one. Parties will of course anticipate this, and will therefore tend to avoid the first agreement in order to wait for a next round. In other words, the *waiting game* is an appealing option.

It is important that arrangements are introduced at the start of the process to stimulate parties to join and to make the waiting game a less appealing option. This will add momentum to the process. The following two examples are an illustration of such arrangements.

#### 3.5.1.1 Early Baseline Date [24]

The highly motivated parties that are willing to commit to agreements at an early stage agree with each other that potential future arrangements in the following rounds will be based on the data that were available at that first moment. Parties that join later will also have to take their measures based upon this early baseline. In any case, measures taken by the early participants will be taken into account in the formulation of later obligations. This arrangement will eliminate an important risk of early participation. What's more, the risk of early participation may become an opportunity. After all, early participants get a chance to frame the problem and to address it in such a way that it is bound to yield solutions that they can agree with. Participants that join later will have to conform to this situation.

There are two sides to this line of reasoning. On the one side, this opportunity may be exploited to such an extent that the early participants twist the situation in their favour and thereby make it virtually impossible for other parties to join later. On the other hand this opportunity represents an incentive for parties to join at an early stage, which is positive from the perspective of the momentum of the process.

#### 3.5.1.2 Voluntary Action Plan [24]

Parties wishing to make an early start emphasize the voluntary nature of the actions to be taken. The plan may be supplemented with elements such as voluntary research and monitoring activities and voluntary pilots. Emphasis on the voluntary nature lowers the threshold to join, and offers early participants an opportunity to introduce and implement their 'hobbies'—which is an extra incentive to join.

### *3.5.2 The Prospect of Gain as an Incentive for Cooperative Behaviour*

The main incentive for progress is the prospect of 'gain'. This means that parties should be convinced that the process is—and will remain—sufficiently appealing to them to participate wholeheartedly and, above all, to bring it to a good and quick conclusion. After all, parties who anticipate a gain will have an interest in concluding the process and collecting their reward.

It is important that the 'gain' for the various parties does not pay off too soon—that is why there is an emphasis on a *prospect* of gain. As soon as a party has received its gain, it no longer has an incentive to be cooperative. At that point there is a risk of opportunistic behaviour: the party in question may withdraw from the process and cease to behave cooperatively. This implies that the process architect should ensure to maximize the chances of gain towards the end of the process.

### *3.5.3 Creating 'Quick Wins'*

Processes always involve a balance between conflicting demands; in this situation, this is obviously the case. On the one hand, parties cannot be expected to commit themselves to subdecisions: nothing is decided until everything is decided. This may create an impression of the process being sluggish, so on the other hand it is important that there are quick wins to be had for the parties. On the one hand the gain should not present itself too soon, for this would stimulate parties to leave the process. On the other hand there should be some pay-off—through quick wins—because parties may also leave the process if the gain is too far away.

### *3.5.4 Ensure that the Process is Heavily Staffed*

Heavy staff means that the participants in the process are the ones who hold high positions within their organizations, and/or have authority in these organizations. There are three arguments for this design principle that indicate that a heavy staff promotes decision making.

The first argument is that a heavy representation promotes the external authority and image of the process. As has been described before, this is an important precondition for progress of the process and the decision making.

The second argument relates to the opportunities that representatives have to stimulate their organization's commitment. They usually have 'commitment power'. If the staff is too light, such commitment power can only be developed by a strong formalization of the relationship between the representative and the represented organization, which in turn will cause a lot of trouble and require discussion and detailed mandates. This may seriously hamper progress. Commitment is much more self-evident in the case of a heavy representation.

The third argument is that a heavy representative may, if necessary, take some distance from the organization he represents. This too is necessary in this kind of processes, in which parties sometimes have to make compromises that may be difficult to accept.

### 3.5.5 Transferring Conflicts to the Periphery of the Process

A process is designed in view of potentially conflicting relationships between the participating parties. A process approach poses an evident, major risk: the parties are brought together, resulting in conflicts being fought out quite fiercely. The metaphor of fighting cocks is sometimes used to illustrate this: a process architect may design a process in order to bring fighting cocks closer together, but once the process has started and the cocks find themselves in a cage, there is a risk that they will start fighting a battle of life and death, more fiercely than ever before.

The process architect will have to take a number of precautions to prevent too many conflicts arising between the parties during the conflict. After all, every process will have a limited tolerance for conflict: too much conflict will affect the mutual relationships to such an extent that it poses a threat to the process.

Process architects may utilize the fact that many processes have a layered organizational structure. There is a core that is enveloped by a number of shells. There may for instance be a structure that involves a steering group, a project group and a working group. The key decisions are made in the steering group. These are prepared in one or more project groups, which may delegate certain day-to-day tasks to a working group. To an outsider this may seem like a jumble of different groups, but it is highly functional from a process point of view, since it offers an opportunity to level potential conflicts. In project groups and working groups, there is not much risk associated with conflict. After all, these conflicts are neither a direct nor an indirect burden for the representatives in the steering group. The positive effects of conflicts—more information, a better overview of contrasting views—may help the steering group members in their decision making.

### 3.5.6 Tolerance Towards Ambiguity

The use of ambiguous terms may stimulate progress in the negotiations. These ambiguous terms will often have a "feel-good" connotation. Examples include terms such as 'quality', 'future-proof', 'efficient', and so on. Leaving the exact meaning of these terms in the open may evoke criticism on the one side: they are vague and it remains unclear what will happen exactly. On the other hand, the use of such terms will allow parties to keep dreaming that their preferences are 'still on the table', which may be sufficient reason for them to keep participating in the process. Such 'constructive ambiguity' allows parties to portray agreements as a victory [9].

The consequence is of course that the process does not end with this agreement. There are three possibilities at this stage. The first is that the process may unexpectedly take an entirely different turn. The ambiguous agreement moves towards the periphery of the process and therefore receives less managerial and media attention. The ambiguous nature is no longer part of the substance of the conflict, and the overall result is that a conflict is avoided that would have proven to be unnecessary anyway.

The second possibility is that the ambiguous agreement remains on the agenda, and will have to be clarified at some point [9]. In this case, too, postponing this explanation may be useful. Parties will retain their commitment to the process at least for a certain period of time. Early clarification could repel parties that are in a losing position, as well as harm the decision making about other issues.

A third possibility is that the ambiguity itself harms the process. In that case, the ambiguity legitimizes parties' loyalty to their chosen course. This may lead to escalation and have negative consequences for the process at a later stage.

Pan Zhongqi qualifies the US policy towards China and Taiwan as constructively ambiguous. During the 1950s, the US—also as a consequence of the Korean war—unambiguously chose to side with Taiwan and against China. In 1954, the US signed the Mutual Defense Treaty with Taiwan authorities to prevent China from reunifying with Taiwan. However, in 1972 US policy took a major turn. The US had returned from the Vietnam War in a weakened state, and needed China's support to keep the Soviet Union under control. The US made a number of important concessions towards China. For instance, it withdrew almost all US troops and military installations from Taiwan, and denounced the Treaty with Taiwan. At the same time, however, Washington committed to an armament programme to arm Taiwan so that it could defend itself. This ambiguous policy, according to the US strategists, would result in neither party risking to start a war. But another result was a serious arms race: neither country wanted to take any risk at all. Perhaps the balance that exists as a result of this ambiguity is too subtle and too fragile to actually persist for a longer time. The US is bound to make a mistake at some point in time, causing the balance to tip towards one particular side—and not without consequences [30].

### *3.5.7 Using Options for Command and Control Created by the Process*

Chapter 2 compared process management to a management style of command and control. It is too simple, however, to assume that there is no role for command and control at all when it comes to process design and management. Certain types of command and control may present parties with an incentive to join the process and act cooperatively; it may thus be a driver for process management. Moreover, parties may become more susceptible to command and control during the process, perhaps because they are in a winning position, or because they learn that consultation alone produces no results. This phenomenon may be useful in process design.

Suppose, for instance, that a minister is hesitant to make a decision about a certain issue because he expects much resistance from the parties involved. This minister may then establish the process agreement that these parties will come to a common conclusion during a process of consultation and negotiation. The next agreement may be that the minister will adopt any consensus reached by the parties, while he will follow his own discretion and make a unilateral decision if parties continue to disagree. In other words, in case of disagreement he will steer by command and control. The threat of command and control may be an incentive for parties to reach consensus. If they do not succeed in this, command and control may be socially acceptable. Those who cannot make a decision in a process will have to accept that others will make the decision for them.

## 3.6 Design Principles that Guarantee the Substance of the Process[4]

### *3.6.1 The Roles of Experts and Stakeholders are Both Bundled and Unbundled*

The above says little about the *substance* of the process. Of course the process architect cannot ignore the substance—even though he is not a substance expert. A process without substance is hollow. At the same time, though, substance can never determine the course of a decision-making process. After all, one of the justifications for a process design is that it is impossible to solve problems on the basis of objective information.

When a decision-making process drifts too far away from the substance, it is vulnerable. It may appear to lack focus. Although parties may have different degrees of tolerance towards the distance between process and substance, there is some sort of *line we dare not cross.*

As a result, the process architect will have to ensure that the process has sufficient substance. The process will have to be designed in a way that allows the relevant substantive ideas to be addressed during the process.

A primary way to guarantee this is of course through openness. If there are many parties, there are many insights. In addition, there should not only be a role for stakeholders in the decision-making process, but also for substantive experts. They can use their substantive knowledge to facilitate the process. They can separate sense from nonsense, conduct sensitivity analyses, familiarize parties with the latest insights, and so on.

---

[4] Refs. [4, 14, 18, 23].

Suppose that government, business and civil society are negotiating about the environmental impact of packages (see also the example in Chap. 2), and there is a conflict about whether the carton box or the glass bottle is more environmentally friendly. The parties stick to their guns: one has data in favour of the carton box, and the other one has data that support the glass bottle. There is, in other words, a stalemate.

This is where there may be a role for an expert. He or she will usually not pass any final judgment. After all, the problem is unstructured and every party will present its own expert. An expert may, however, introduce insights that help break the stalemate, and thereby raise the substantive level of the discussion. The expert can point out the innovation potential associated with packages: some may be optimized through environmental technology, others may not. The expert can underline how a package's environmental profile depends on certain parts of the package. The profile of a carton box may for instance be influenced by the plastic cap, and that of the glass bottle by the lid. There may be potential innovations that can shift the discussion. Opponents of the carton box may be able to live with the carton box if the plastic cap is eliminated, and if its improvement potential is used more efficiently. This kind of new information may break a stalemate—after all, the discussion starts moving again—and may lead to better substantive decision making.

There is another reason why expert involvement in the process may be important. It offers stakeholders the opportunity to question experts about the scientific level of their research results or beliefs: which assumptions are they based on, which data have been used, which system boundaries have been established, and so on.

Perhaps there are researchers who, based on their analyses, argue that the polyethylene bag has the best environmental profile. Stakeholders are often knowledgeable, and may fire a large number of questions at the experts. Which data have been used? Do the researchers realize that the bags often rupture and that product is lost, and is this environmental damage taken into account? Or do the researchers know that these bags are only produced at a few locations in Europe, and that the use of these bags therefore always involves extra transport? And so on.

### *3.6.2 From Substantive Variety to Selection*

A second important guarantee for sufficient substance of a process results from the principle of variety and selection. This means that a large number of substantive insights and ideas may be introduced at the start of a process and that in the end, some of these insights and ideas will be selected from among this variety.

Tolerance towards much variety at the start of a process ensures that all relevant insights and ideas have a chance to be addressed during the process.

First of all, this increases the substantive quality of the decision making. If parties limit their attention to certain insights and ideas too early in the process, this may affect the quality of the decision making. If a variety of insights and ideas is taken into account, this makes it harder for parties to call into question the selection of insights and ideas at a later stage.

Suppose a country is in need of a location for a new airport. Some parties suggest constructing this airport in the sea. Other parties see no sense whatsoever in this idea, or they are convinced that it is financially and technically unattainable.

The principle of variety and selection implies that it may be sensible to take this option into account anyway. Suppose it is evaluated anyway and it turns out that it is indeed financially and technically unattainable. From the opponents' perspective, this evaluation may be a loss of time. But is it really? When another location is selected during the process, and the supporters of an airport at sea have difficulty accepting this other location, the earlier evaluation does have significance. These parties cannot claim that the option of an airport at sea has not been evaluated, and then use this as an argument to stall the decision making about the other location. In addition, it will be obvious that the quality of the decision making benefits from evaluations of all options. Parties that vehemently reject an airport at sea may after all be wrong, for instance if research shows that this is in fact a viable option as long as the airport is not too far away from the shore.

### 3.6.3 The Role of Expertise in the Process

In short, there should be a role for knowledge and expertise in the process. How can this be achieved? It is too simple to suggest that science should just be 'dragged into the process'. This would ignore the fact that if science is to play a role in such a process, it will operate outside of its actual domain.

Traditional science generates its own research questions. Those are questions that emanate from the development of the scientific discipline in question and that are, rationally, next in line to be answered. The kind of answers needed in the process, however, does not stem from science, but rather from society. Science is more or less forced to reach a conclusion about a question that is not yet at the top of the scientific agenda. This implies, among other things, that science cannot formulate a definitive, certain answer to the question at hand. The reaction from science will always harbour some uncertainty, and there will remain room for several answers from science [11, 21]. Some of this room will, more or less implicitly, be filled by the values of the researcher and his organization. Put differently, whereas science used to limit itself to those domains that are characterized by 'hard facts and soft values', these processes will now force science to reach conclusions about issues that are characterized by 'soft facts and hard values' [6, 10]. This will lead to discussions between scientists that can no longer be settled by additional research. There will always be differences of opinion at the level of values.

This gives rise to the following question. Taking all of this in mind, how can process arrangements be shaped in such a way that experts and scientists, despite their differences in opinion, can make an optimal contribution to the progress of the process and to its quality and the resulting decisions? The answer is: by shaping the discussion between the scientists as a process as well. Scientific assessments turn out to be more effective if they are shaped like processes [7, 28].

The processes in which scientists and experts operate may be meaningful at two levels. At the first level, the interaction takes place between scientists. Together, scientists identify the areas of agreement and disagreement, respectively. They publicize the facts and causalities that they agree about, while they aim to minimize the number of issues that they disagree about. They do this by designing shared research protocols that can contribute to the generation of new knowledge that will reduce their differences of opinion. Another possibility is that even though the scientists will not agree with each other, they will be able to reach consensus about the margins within which they disagree. Remaining disagreements can be managed for instance by allowing dissenting opinions, the establishment of competing assessment processes and the inclusion of minority reports [25].

At the second level, there is interaction between the scientists and the decision makers [19]. This is the level at which there is actual bundling between decision makers and experts, and at which the envisioned quality of the decision-making process is effectuated.

> 'The concrete development of such processes at these two levels is illustrated by the example of the Intergovernmental Panel on Climate Change (IPCC). In December 1988, the United Nations (UN) General Assembly unanimously passed a special resolution calling for the adoption of a 'framework convention' on climate change. In line with his charge, the United Nations Environment Programme (UNEP) and the World Meteorological Organization (WMO) set up the IPCC to undertake a comprehensive review of the area and to make recommendations [24]. In order to maximize the authority of the IPCC reports, an extensive review process has been designed. The review process took place in two rounds. First, drafts were circulated among specialists in the relevant areas. In the second round, the revised drafts were distributed among governments. As a rule, governments sent these drafts to ministry officials, to scientists and individuals at the boundary between science and policy.
>
> Finally, the lead authors had to include the comments into a final draft that was submitted for acceptance to the working group plenary meeting. While the lengthy chapters in the bulk of the IPCC reports only require the acceptance by the working group, the shorter and more focused executive summaries and the summaries for policy makers had to be approved line-by-line by the IPCC plenary consisting of government officials' [25].
>
> 'Most of the dispute in the plenary sessions revolved around the question of what has to be included in the summaries and what not. Due to the consensus principle all delegates have to agree to the final wording. Opposing positions have to be articulated and explained in the plenary session and if no compromise between opposing positions can be found, the discussion will be continued in smaller contact groups. Although this mechanism in most cases delivers acceptable solutions, sometimes certain countries try to push their claims even further. In this case when absolutely no compromise could be reached in the small groups, a dissenting vote will be included in the text naming the dissenter. Since this dissent is made public through this procedure, countries usually dislike to fall back on this option—especially because it is mostly the same small number of countries with clear political or economic interest… that try to weaken certain statements…' [25].

The design principles presented here may help the process architect in designing the process. Chapter 4 will describe which activities are needed for this. In Part III, we will discuss each of these design principles in further detail. There we will not discuss them from the perspective of the process architect, as we did in this part, but from that of the process manager.

# References


1. Arnstein SR (1971) Eight rungs on the ladder of citizen participation. In: Edgar en SC, Passet BA (eds) Citizen participation: effecting community change. Praeger, New York, pp 69–91
2. Bohman J (1996) Public deliberation: pluralism, complexity and democracy. MIT Press, Cambridge
3. De Bruijn JA (2000) Processen van verandering. Lemma, Utrecht
4. De Jong WM (1999) Institutional transplantation: how to adopt good transport infrastructure decision-making ideas from other countries. Eburon, Delft
5. Dixit A, Nalebuff BJ (1991) Thinking strategically. The competitive edge in business politics and every day lifes. Norton, New York
6. Dutch Scientific Council for Government Policy (2008) Onzekere Veiligheid, verantwoordelijkheden rond fysieke veiligheid. Amsterdam University Press, Amsterdam
7. Farrell A, VanDeveer SD, Jäger J (2001) Environmental assessments: four under-appreciated elements of design. Glob Environ Chang 11(4):311–333
8. Fischer F (2000) Citizens, experts and the environment. Duke University Press, Durham
9. Fischhendler I (2004) Legal and institutional adaptation to climate uncertainty: a study of international rivers. Water Policy 6(4):281–302
10. Funtowicz SO, Ravetz JR (1993) Science for the post-normal age. In Futures 25(7):735–755
11. Gibbons M, Limoges C, Nowotny H, Schwartzman S, Scott P, Trow M (1994) The new production of knowledge: the dynamics of science and research in contemporary societies. Sage, London
12. Guba EG, Lincoln YS (1989) Fourth Generation Evaluation. Sage, Newbury Park
13. Innes JE (1996) Planning through consensus building: a new view of the comprehensive planning ideal. J Am Plan Assoc 62(4):460–472
14. Jasanoff S (1990) The fifth branch: science advices as policy managers. Harvard University Press, Cambridge
15. Kheel TW, Lurie WL (1999) The keys to conflict resolution: proven methods of settling disputes voluntarily. Four Walls Eight Windows, New York
16. Kiser L, Ostrom E (1982) The three worlds of action: a meta-theoretical synthesis of institutional approaches. In: Ostrom E (ed) Strategies of political inquiry. Sage Publications, Beverly Hills, pp 174–222
17. Mayer IS (1997) Debating technologies. A methodological contribution to the design and evaluation of participatory policy analysis. Tilburg University Press, Tilburg
18. Miranda ML, Miller JN, Jacobs TL (1996) Informing policymakers and the public in landfill siting processes. In: Technical expertise and public decisions. Institute of Electrical and Electronic Engineers, Princeton
19. Mitchell R, Clark W, Cash DW, Alcock F (2002) 'Information as Influence: How Institutions Mediate the Impact of Scientific Assessments on Global Environmental Affairs', Faculty Research Working Paper 02–044. Kennedy School of Government, Harvard University, Cambridge
20. Moore CW (1996) The mediation in process: practical strategies for resolving conflict. Jossey-Bass, San Francisco
21. Nowotny H, Scott P, Gibbons M (2001) Rethinking science: knowledge and the public in an age of uncertainty. Polity Press, Cambridge
22. Raiffa H (1982) The art and science of negotiation. Belknap Press, Cambridge
23. Science and Public Policy (1999) Special issue on scientific expertise and political accountability, vol 26, no. 3
24. Sebenius JK (1991) Designing negotiations toward a new regime. The case of global warming. Int Security 15(4):110–148
25. Siebenhüner B (2003) The changing role of nation states in international environmental assessments-the case of the IPCC. Glob Environ Chang 13(2):113–123
26. Sparks A (1995) Tomorrow is another country: the inside story of South Africa's negotiated revolution. Struik, Sandton

27. Stern PC, Fineberg HV (eds) (1996) Understanding risk informing decisions in the democratic society. National Academy Press, Washington, DC
28. Ten Heuvelhof EF, Nauta C (1997) Environmental impact; the effects of environmental impact assessment. Project Appraisal 12(1):25–30
29. Watkins MD (2007) Teaching students to shape the game: negotiation architecture and the design of manageably dynamic simulations. Negotiation J 23(3):333–342
30. Zhongqi P (2001) The dilemma of deterrence: US strategic ambiguity policy and its implications for the taiwan strait. The Henry L. Stimson Center

# Chapter 4
# The Process Architect in Action: Making a Process Design

## 4.1 Introduction

The design principles outlined in the previous chapter may be helpful when making process agreements, but of course the main question remains how such agreements are made. In this chapter we will answer this question as follows. (Both the current chapter and Chap. 5 are partly based on actual process designs and draft process designs that we made in the past[1]).

In Sect. 4.2 we will indicate that process agreements are always the result of some kind of negotiation. It is almost impossible to standardize process agreements, because the configuration of actors and the substance of the issue at hand will differ from situation to situation. In Sect. 4.3 we will discuss an important condition for process management: the parties involved should have some sense of urgency regarding the fact that they need each other to solve a particular problem—and that a process is needed to do so. If there is no such sense of urgency, process management is unlikely to succeed. Section 4.4 provides an overview of the main activities that a process architect undertakes and that result in a process design.

## 4.2 The Process Design as a Result of Negotiation

An important condition for the success of a process design is that it should be appealing to each of the parties involved: they should be convinced that the design offers them a fair chance of influencing the decision making and that it will not harm their core values.

It will be difficult for any one party (or for an independent third party) to draw up an appealing process design unilaterally, especially if there are significant

[1] See [2–13].

H. de Bruijn et al., *Process Management*, DOI: 10.1007/978-3-642-13941-3_4, 

conflicts of interests between the parties. The conclusion is obvious: an attractive process design, which all parties have ownership over, can only originate if these parties can participate in shaping it. In other words, the process design itself is one of the outcomes of a process.

At first glance, negotiations about a process design may seem to be inefficient—particularly in the eyes of an initiator who initially thought he could implement a substantive change unilaterally. After all, this initiator (1) starts out with a substantive initiative, and has to accept the fact that he depends on others and needs to enter into (2) a process of discussion and negotiation with these other parties. Then it turns out that (3) even the rules of designing the process are subject to negotiation. This is quite different from the original idea to implement a substantive initiative: moving from substance to a process that will lead to process agreements. However, the same side note that we made earlier should be made here as well: the drawing up of a process design can be a time-consuming and major negotiation game—for instance when it comes to peace conferences, which are characterized by negotiations about the process itself, as we have seen. Making a process design can also be a more implicit activity. Imagine for instance a manager who, prior to a change, consults with his stakeholders about a limited set of process agreements: who is involved in the decision making, and when and how? The principle, however, is the same: process agreements are the result of a process. Which are the positive effects of this?

- Of course process agreements are more likely to succeed when they are a product that enjoys shared ownership of the parties involved. If this is not the case, it is relatively easy for a party to distance itself from the process agreements during the substantive negotiations if it feels that these process agreements favour the other parties. Negotiating about processes means that parties can learn from each other: about their respective interests, sensitivities, core values, space for solutions, etcetera. If this learning process were to take place during the substantive negotiations, these would be seriously disrupted. Negotiations about process agreements thus have the advantage that this learning process can take place during the procedural negotiations without such a disruption.
- Negotiating about a process design often implies that the substantive negotiations have started as well. After all, the process agreements proposed by parties are often inspired by the substantive issues they wish to address. In this case, too, these substantive issues will be addressed, but without the parties needing to reach actual substantive agreements. This allows them to learn about the substantive agenda of the negotiations.
- Parties also learn about the need for process agreements. When they notice during the procedural negotiations that there are strong differences between the parties' interests, that agreement is not self-evident, and that the relationship between the parties generates conflict, there is an increased chance that they will internalize the process agreements and accept the idea that every party should respect these. This is an important notion as well, because during the phase of

substantive negotiations (in other words, when there is a set of process agreements), parties may be tempted to ignore process agreements at moments when it suits them better from a substantive point of view.

## 4.3 The Need for a Sense of Urgency

Process management can only succeed if there is a sense of urgency among the main stakeholders [15]. Some sense of urgency is often a prerequisite for a change process, but in this case the sense of urgency comprises at least two components:

- A substantive component: parties should be convinced that there is an issue that needs to be solved.
- A process-oriented component: parties should be convinced that this issue can only be solved through cooperation in a process.

Without this double sense of urgency, parties will not easily be prepared to negotiate about process agreements. When parties only have a substantive sense of urgency, and try to solve an issue, one of two things may happen. Either they succeed, perhaps unexpectedly, to solve the issue—or they do not, which will eventually lead to the sense of urgency regarding the fact that a process is needed.

Two major European rivers, Maas and Rhine, flow into the North Sea in the southwestern part of The Netherlands. The Netherlands is protected from these rivers by dikes. For many centuries, these rivers have been conquered with increasing success by continuously raising and reinforcing the dikes. However, experts believe that this strategy cannot go on forever. They hold that climate change will lead to an increasing volatility of the water level in these rivers, and that increasing water levels are to be expected. In addition, the low-lying areas in the western part of the country, on the land side of the dikes, are 'setting', meaning that their level is gradually dropping. Last but not least, this same climate change may result in a considerable sea-level rise, which would make the discharge of river water into the sea increasingly difficult. Focusing exclusively on raising the dikes is therefore no longer an option, according to experts. What is needed is a completely different mindset about the protection against high river water levels. What is needed is a paradigm shift: from protection by dikes towards 'more space for rivers'.

Such a paradigm shift will not occur without problems. From the perspective of local governments, water authorities and citizens who live near the river, this will not always imply an improvement. They will have to sacrifice space, and this is not always appreciated. A paradigm shift will be difficult to accomplish in such a context. Opponents will point out that raising the dikes has been a wise strategy for many decades, if not centuries, and that there is little reason to believe that this is the moment to change this strategy. *Never change a winning strategy* is what these opponents will argue.

Nevertheless, during a process that lasted around ten years, this new paradigm has become generally accepted in The Netherlands, and the preparation of some forty concrete 'More space for rivers' projects has been initiated—with broad political support. A factor that certainly contributed to this is the fact that two near-floods occurred in the Dutch river region in the mid-1990s. The danger was so significant and real that large-scale, dramatic evacuations were carried out. Terror was widespread, and the resulting sense of urgency undoubtedly increased the possibilities for change.

This is 'only' a substantive sense of urgency—which is not sufficient for a process to be successful. Some local authorities could, for instance, shirk the decision making, or obstruct it, driven by a NIMBY-like attitude. Such behaviour might prove to be effective: the costs of the rising water level might well be borne by others that act more cooperatively. How, then, can a process-oriented sense of urgency be created? How can local authorities, water authorities, provinces and the national government be convinced that cooperation and a proper process are needed to solve this problem?

The government has (1) identified 700 potential measures that can help solve this problem; (2) out of these 700 measures, only a few dozen are necessary to solve the problem; and (3) there is room for compensation. This was a substantive action, but it also created a sense of urgency regarding the need for cooperation in a process. After all, these measures are interrelated and they touch upon the interests of all stakeholders. Stakeholders who, given these 700 measures, refrain from participating in the process may well harm their own interests. Chances are that others will make the decisions for them. There is often a contagious effect when it comes to processes: the more players participate, the riskier it becomes for others hold aloof.

Two things need to be noted in relation to the process-oriented sense of urgency. Firstly, one or more parties may not recognize the need to negotiate. They may have a strong BATNA—'the Best Alternative to a Negotiated Agreement' [14]. Suppose a strong BATNA is held by the actor who is the only one who has the necessary means (money, information, legal authority and so on). In that case it is to be expected that this actor will be unlikely to make concessions, which will hamper the progress of the process. In other words, his *process sense of urgency* is low. The process architect may solve this problem by shaping the agenda of the process in such a way that this actor has a prospect of gain that greatly surpasses his BATNA [1].

Secondly, a process approach can be applied too early in a decision-making process. A consultant may note, for instance, that the large number of stakeholders involved in a particular issue calls for a process approach. This is often a wrong kind of efficiency: first, the parties need to develop a sense of urgency that they will be unable to solve the problem without a process. This usually takes some time.

Decision making in a network is characterized by natural dynamics: there is a substantive initiative, and there will be a process of haggling and daggling between the actors that will either produce a result at some stage, or stagnate. In the latter case, parties may develop a sense of urgency that will allow them to make process agreements. Sometimes the best advice may be to let the decision-making process take its course for a while, until the parties become convinced that there will be no progress unless they cooperate. For the process architect, it is better to be too late than too early. If a process architect enters the scene too early—when there is no process sense of urgency yet—there is a risk that the process will stagnate. If he enters the scene late, on the other hand, this is an advantage: the sense of urgency has grown, making parties more willing to invest in a process.

If the process is initiated too early, chances are that it will peter out: parties are unable to make process agreements, or, if they do make such arrangements, they treat them carelessly. After all, they do not have a sense of urgency that a process is needed in the first place. This will lead to a risky situation. If parties' increasing awareness does eventually lead to a sense of urgency, they will have to start cooperating in a process after all. There will, however, be a *burden of the past*: parties remember negative experiences with an earlier process, which may hamper cooperation in the process that is now needed. In other words, organizing processes too early or without due consideration may cause damage to future processes.

## 4.4 The Process Architect in Action: Designing a Process

This paragraph will describe the activities of a process architect that result in a process design. We will start with a few introductory remarks.

Firstly, this will be a complete overview of possible activities. However, it is not always necessary that all of these activities are undertaken in order to arrive at a good process design. Depending on the substantive complexity and the nature of the conflicts of interest, the number of necessary activities will vary. Under some circumstances it may be opportune to skip activities, or to implement them to a limited degree.

Secondly, the order of activities is less imperative than suggested by the list below. Sometimes another order is opportune, sometimes iterations are needed and activities are repeated one or several times.

Thirdly, a process architect who has a sense of managerial relationships will realize that it may be wise not to make all activities explicit. We will illustrate this with respect to the question how a process manager should structure an agenda.

Lastly, it should be pointed out that going through these activities, in combination with a correct application of the designing principles, does not necessarily result in a good process design. Designing a process is not a mechanistic activity. A good process design requires managerial creativity and sensitivity. There is a touch of virtuosity about some designs. The process architect also needs to have an eye for the elegance of process arrangements, he should have a sense of managerial relations, he should know about managerial customs, about the burden of the past and about people and characters. In addition, superior language skills and well-developed conceptual talent are indispensable to achieve a process design that commands managerial support.

Figure 4.1 describes the activities a process architect performs when making a process design.

### *4.4.1 Exploring the Problem Together with the Commissioning Party*

Although the essence of a process may be that it involves several actors, the start of a process tends to be a matter of one actor or a limited number of actors. This initiator or these initiators contact the process architect and indicate that there is an

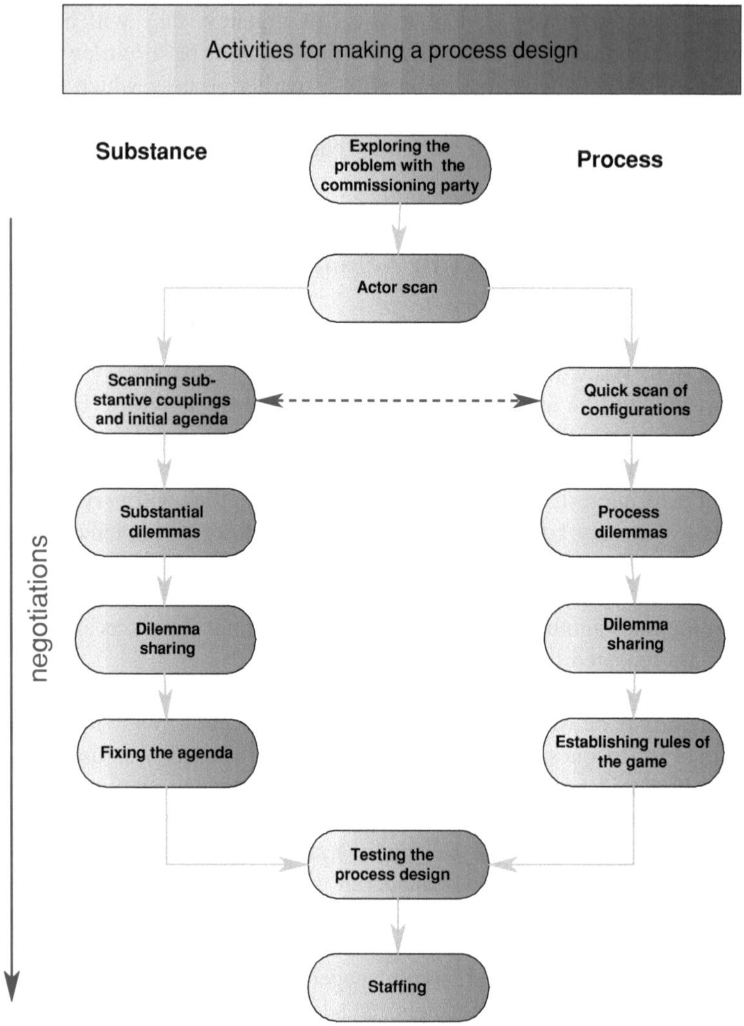


**Fig. 4.1** Activities for making a process design

issue they wish to realize, and that they are concerned about the course of the decision-making process so far. They feel, for instance, that the quality of the decisions is low and progress is slow, or they are concerned about the level of support for the decisions or about the level of managerial commitment. The initiator then commissions the process architect to get the decision-making process started. In the initial phase, he will perform the following activities.

First of all, he conducts an initial problem exploration in consultation with the commissioning party. Inevitably, this exploration is rather biased towards the views of the commissioning party. The process architect should emphasize that this problem exploration is no more than a first exploration and that if a process is to be successful, all relevant actors should endorse the definitive problem definitions and agenda.

Secondly, the commissioning party and the process architect draw up an initial list of relevant actors.

The third step in this initial exploration together with the commissioning party is to map the sense of urgency of the process. The commissioning party will almost certainly feel a certain urge to advance the decision making. Otherwise he would probably not have contacted the process architect. But is the same thing true for the other actors? Do they also feel the need to address the decision making, or do they feel no such urge, or do they in fact perceive the stagnation as positive? This is where the process architect should make an inventory of the views of the other actors.

If the process architect and the commissioning party conclude that the sense of urgency is uncertain, they have the following options:

- Their first option is that they jointly decide that under these circumstances, designing a process stands no chance of success. If a process is started anyway, it will either not get off the ground or fail, which may harm relations between actors. This may place a heavy burden on any following process.
- Consequently, the second option is that the commissioning party and the process architect decide to postpone answering the question about the sense of urgency, and explore the actors' actual sense of urgency during a following round. A sense of urgency may for instance emerge during the discussion of the agenda, which contains more items than just those suggested by the commissioning party.
- A third option is that the commissioning party and the process architect explore the possibilities of increasing the pressure on the process, thereby creating a sense of urgency among the other actors as well.

### 4.4.2 Actor Scan

As soon as the initial exploration with the commissioning party has been completed, there will be a preliminary scan of the relevant actors. The first selection is made in consultation between the commissioning party and the process architect. The list will be expanded in the course of the scan. During the analysis of written documents and during the interviews with various actors, new potential actors will emerge who were not initially proposed by the commissioning party, but who are nevertheless important for the progress and quality of the process. After consultation with the commissioning party, the process architect may involve these actors in his scan as well.

The aim of the scan is to gather certain data about each individual actor. For each actor, the process architect will collect information about views, interests and core values; about risks and opportunities they may identify once the process is started; about incentives and disincentives; and about pluriformity. The following section will briefly explain these points.

#### 4.4.2.1 Views, Interests and Core Values

The process architect maps each actor's points of view about a particular issue. Does the actor in question support or oppose the views of the commissioning party? Or is the actor keeping his options open? Does the actor mention conditions that must be met before he will cooperate? How passionate is his attitude towards the proposal?

The process architect discovers interests by gathering information about the background of the actor's point of view. Why is he a supporter or an opponent? What does he expect to accomplish? In practice, it is usually not the actor himself who clarifies his interests, but rather the other actors: they are in a better position to explain why an actor takes a particular view.

When clarifying interests, it is sensible to make a distinction between formal and informal interests. The actor himself may identify the formal interest behind his views. This interest refers to his organization's 'raison d'être'. Informal interests are usually much harder to describe, and they may even have a slightly vulgar nature. This, however, does not make these interests less relevant to the process.

Once the process architect has gained an impression of the views and interests of an actor, he can get an initial idea of their 'elasticity'. Which are the potential changes in this point of view, given the actor's interests?

Interests that are special and particularly important to the actor can be regarded as core values. Core values do not exclusively relate to an issue pertinent to the decision-making arena; they have a more generic nature. They are, however, of crucial importance to the actor, and also apply to the issue at hand. Core values should be revealed, because they require protection in special arrangements. After all, if such protection is offered, it is easier for actors to join a process (as will be further explained in Chap. 6).

#### 4.4.2.2 Risks and Opportunities

The process architect maps the opportunities identified by each of the actors with regard to the issue the initiator wishes to address during the process. It is important that the process architect finds out which other issues can be coupled to this issue in order to make the process appealing to the other parties. The prospect of a process with plentiful opportunities may stimulate parties to commit to this process. The process architect will also address potential threats from the actor's perspective: which kind of agenda would make an actor hesitant to join the process?

#### 4.4.2.3 Incentives and Disincentives

The process architect should also map the incentives and disincentives for each actor. Incentives are the factors that stimulate an actor to show proactive behaviour; disincentives are the factors evoking reactive behaviour or even obstruction.

#### 4.4.2.4 Pluriformity

Finally, the process architect will map the pluriformity of the participating actors. Is the total of actors homogenous or are there major differences between them? Does a party representative speak on behalf of this party or is this party itself pluriform to an extent that the representative does not speak for his entire party? A clear picture of this pluriformity provides the process architect with an idea of the stability of an actor's points of view. After all, an actor who represents pluriformity may be less stable with regard to his viewpoints than an actor whose view is uniform.

There are three potential sources for such scans per individual actor: an analysis of the written documents, an interview with the actor himself, and interviews with other actors about the reputation of the individual actor in question.

#### 4.4.2.5 An Actor Scan is a Continuous Activity

An additional remark should be made about actor scans. Many organizations are quite familiar with actor scans. Organizations that realize that they need other actors in order to achieve their goal, will also realize that such a scan is indispensable for making progress. They will perform such a scan—which is often referred to as a field-of-force analysis. The result is a list of actors, their views and their resources, telling the initiator which actors to influence if he is to achieve any result.

Such field-of-force analyses tend to be once-only activities, which are carried out when the initiator becomes aware of his dependencies. These field-of-force analyses are driven by the idea that the initiator wishes to achieve something and that other actors have to be stimulated to cooperate—or, at any rate, not to use their obstructive power. From this perspective, actors represent barriers that have to be overcome, and the field-of-force analysis is a once-only means of achieving this.

The actor scan as part of a process design has a somewhat less instrumental nature. Rather than being a once-only activity, the actor scan should take place continuously, embedded in the process. The reason for this is twofold: interests cannot be fully known, and interests as well as views may change in the course of the process, for example under the influence of the process itself.

Unlike the field-of-force analysis, the continuous actor scan aims to disclose views, interests and incentives in order to organize a process that will result in an outcome that does justice to a maximum number of interests. Actors therefore actively participate in the process. One of the consequences of this active participation is that actors undergo learning processes. They become aware of other actors' views, participate in research trajectories, communicate intensively with other actors in the process, and so on. They will be influenced by this information and these networks. Their views will evolve and they will learn what they consider to be their actual interests. In short, the views and interests of actors who participate intensively in a process will gradually evolve. As a result, however, the outcome of the actor scan will be obsolete after a while. In other words, an actor scan requires updating, partly because of the process itself. If the outcomes of an actor scan remain unchanged for a longer period of time, this may be a sign that a process is not developing properly.

### *4.4.3 Quick Scan Configurations*[2]

The information gathered will enable the process manager to make a quick scan of the configuration of the parties. Taking the initiator's issue as a starting point, the process architect composes an overview of the views and interaction patterns of the actors involved.

The analysis will reveal which actors hold relatively extreme views and which actors propagate views that meet broad support. It also becomes clear which actors communicate frequently and intensively and which actors operate in social isolation (see Fig. 4.2).

Actors 1, 2 and 3 are marginal. They hold extreme views—which do not necessarily correspond with each other—and they do not communicate frequently. Actors 8, 9, 10 and 11 occupy a central position in the network. Their views approach the average view, and they communicate frequently and intensively.

In the first place, this analysis shows which actors occupy a key position in the network from a substantive and communicative perspective. They are the actors who have a well-developed network of connections and who hold views that resemble those held by others. For this kind of actors, cooperation in a process is usually self-evident: process agreements tend to codify a kind of behaviour that is natural to them anyway.

Secondly, the analysis shows which actors are marginal. These are actors who hold extreme views, and, more importantly, who maintain relatively few relations.

The analysis may be expanded by an inventory of the distribution of resources among different actors. If the actors in the top right hand corner of the figure (actors 8, 9, 10 and 11) are the same actors who have many resources at their disposal, the process will take its course. In this case, the actors with intensive communication and average and widely supported views also have resources. They can therefore be expected to communicate properly about the use of their resources, and to use those in areas that enjoy wide support. The situation is more

2 Ref. [16].

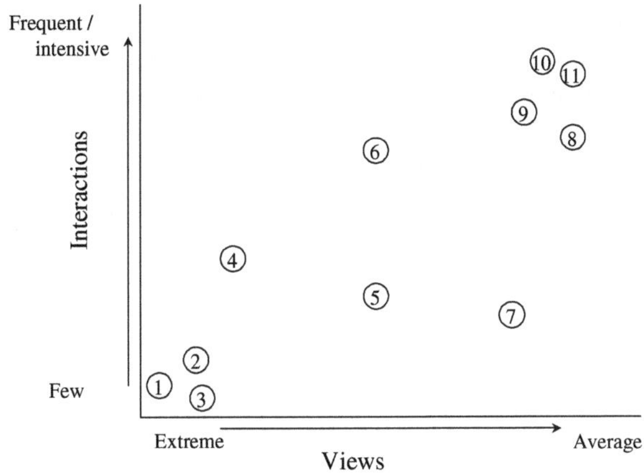


**Fig. 4.2** The position of actors in a network

problematic if the resources are limited to actors who are in a marginal position. It is of major importance to the process that they remain committed to it, but given their marginal position this is not self-evident. There are several ways to commit them anyway:

- reframing and renaming the agenda of the process, allowing them to recognize more of themselves in the process and thus encouraging them to communicate more intensively;
- coupling the process to other processes and issues regarding which they hold more average positions and also maintain more relations.

The actor scan will produce an overview of issues that are relevant to the process. The process architect composes a list of these issues, thereby making a distinction between substantive and process-related issues. This will be outlined below.

## *4.4.4 Scan of the Substantive Couplings, and the Initial Agenda*

While scanning the actors and analyzing the configuration, the process architect will gain an impression of the issues that make a process appealing to the parties. The result is an image of the potential substantive couplings between issues during the process. Based on the scan of these substantive couplings, the process architect will formulate the agenda that constitutes the start of the process.

### 4.4.4.1 Converting Issues and Interests into Dilemmas and 'Dilemma Sharing'

When drawing up the agenda, the process architect is faced with various substantive views of the parties, many of which will be conflicting. Opposite views are a source of conflict. How should the process architect deal with this? The answer is: by framing issues as a dilemma where possible. We define a dilemma as a problem to which there are two opposite solutions, which both have their pros and cons [17]. There is not one single, obvious, exclusive solution, but there are several, each with their strengths and weaknesses. Framing a conflict as a dilemma affects the process in a number of positive ways.

Suppose a major company undertakes a limited number of international activities. While these activities are quite visible to media and consumers, they generate substantially less revenue than the national activities. The international activities are carried out by a small group of employees who are very passionate supporters of these activities. They admit that these activities are perhaps less profitable than the national activities, but, as they point out immediately, the national market is more and more demand-controlled, resulting in competition based on price alone. That is not a very appealing future. The future abroad is much more promising, which is why they feel there is a need for investments in the international activities.

Others, however, only have an eye for the concrete statistics. They feel that the profits made in the national market are used to subsidize the international activities of a limited number of enthusiasts.

The company's management aims to develop a new mission, and values a certain level of support. After all, the organization comprises many highly trained and scarce professionals, and unilateral managerial choices may stimulate them to leave. In the discussion about the mission, there is an escalating conflict about whether or not the international activities should be part of the mission, and if yes, how prominently.

The solution to this conflict may lie in framing this conflict as a dilemma: national activities are profitable, but they will be subject to a demand-controlled market and will thus become less profitable; international activities are less profitable, but their future market is indeed appealing. A working group is established with supporters and opponents of the international activities and with a few neutral members. This working group is charged with elaborating both potential missions. For both missions, it is to map the pros and cons, where possible in a fact-based manner. Both parties are thus offered an opportunity to elaborate their preference, while they also have room to indicate the disadvantages of the other mission. They are, however, forced to produce a fact-based and proper description of both alternatives. The resulting process may lead to parties developing more sympathy for each other's arguments and views, and to some arguments and variants disappearing off the table because they cannot be reasonably and factually supported. The result could be something along those lines: there will be international activities, but only in countries where reasonable long-term revenues are to be expected with a relatively high degree of certainty. Moreover, these activities are to be monitored biannually.

Which are the advantages of framing a problem as a dilemma?

- Framing opposing views as a dilemma relieves the architect of the need to make substantive choices; the parties in question will note that their views are recognized by the process architect—but also that they are part of a dilemma.
- When the parties in question recognize that opposing views constitute a dilemma, this reduces conflict. When opposing views are a dilemma, the question is no longer which view should prevail, but how to reconcile both views. This forces parties to consider a trade-off between both views.

- If there is a dilemma, parties will inevitably put their own views into perspective. After all, a dilemma implies that both views, although opposing, may be correct.
- If parties are able to put their own views into perspective, they become more susceptible to process agreements. After all, these aim to produce a trade-off between a dilemma's extremes as a discussion outcome.

It will be clear that this kind of dilemma sharing can only be functional if parties recognize themselves sufficiently in the dilemma. Constructing dilemmas is therefore an activity that should be shared between the process architect and the parties in the process. Parties are often engaged in a learning process as a result of dilemma sharing: first they formulate their own views, then they learn that these are not unambiguous (because they are part of a dilemma), and that other views may therefore also be justified and reasonable. The process of constructing and sharing dilemmas when designing process agreements is thus an important basis for successful process management.

### *4.4.5 Substantive Dilemmas and Establishing the Agenda*

The process architect classifies the substantive dilemmas, indicating how they relate to each other substantively. For instance, a choice regarding a particular dilemma may strongly influence the options regarding another dilemma.

The classification may be shaped like a kind of decision tree, but the dilemmas may also be grouped in a number of clusters. The latter option creates the outlines of a package deal.

The process architect also proposes the order in which the dilemmas will be dealt with during the process. He then establishes the substance of the agenda.

When drawing up the agenda, it is important for the process architect to assess different possible ways to deal with each of the dilemmas. Some will be easy to solve, others will be more complicated. When establishing the agenda, he may aim to order the dilemmas in a way that is conducive to a proper process flow. This ordering may imply that he spreads the discussion of the dilemmas intelligently over a certain period of time, or that he couples particular dilemmas while decoupling others.

The following strategies will help a process architect to deal with dilemmas. In some of these cases, a dilemma is either resolved or neutralized; in other words, a choice is no longer necessary, or it can be postponed.

#### 4.4.5.1 Resolving the Dilemma by Synthesis

The process architect foresees the possibility that the parties will fully agree with each other. The solution is new and fully meets the parties' interests. He therefore reserves time during the process for actors to discuss with each other—and if needed with third parties—the views that form the opposites of the dilemma.

A well-known example is the following: two parties both claim a shipment of oranges. The conflict seems insoluble, as both parties claim exactly the same thing, and whatever one of them receives will be deducted from the part the other party gets. At some stage, however, the discussion shifts from the points of view to the interests of those parties. The one party's interest, it turns out, is the continuity of its soft-drink factory. It needs the oranges for their juice. The other party has an entirely different interest. It makes perfumes and therefore needs the orange peels. Now that the two parties are aware of each other's interests, they are able to solve the conflict in a most elegant way. The solution is obvious in this case. One party receives the flesh of all the oranges, and the other gets all the peels. Both parties are completely satisfied, because their interests have been taken into account. Had the discussion not been conducted at the level of the underlying interests, the parties might have reached a compromise of splitting the shipment in two equal parts. By negotiating at the level of interests, on the other hand, the parties have achieved a much better outcome.

#### 4.4.5.2 Pilot Option

The parties decide to implement one of the options fully, while implementing the other one at a modest scale, as a pilot. The actors organize a process in which they monitor the results of the pilot jointly. If the pilot produces positive results, the parties will reconsider their earlier decision.

#### 4.4.5.3 Mothball Variant

The parties choose one of the options and decide to keep the other option mothballed—in other words, available as a back-up where possible. This implies that resources are allocated to maintaining the knowledge about this option, and that—if relevant—the option is not made definitively impossible in terms of space allocation, and so on. If required, the variant rejected in the decision making can be de-mothballed immediately.

#### 4.4.5.4 Developing Options in Parallel

Parties agree to investigate whether it is possible to implement both proposals simultaneously and in parallel. The parties examine until which point in time this will be possible, and which conditions must be met in order to allow for parallel development of the two proposals. The parties agree on a monitoring process that will produce data to discuss the dilemma again later, but then on the basis of information that is richer than what is available now, and that was gathered jointly.

#### 4.4.5.5 Growth Model

Parties explore whether it is possible that the one option develops into the other option during the course of the process.

During the discussion about expanding the Maasvlakte area in the Rotterdam Port, several spatial models were debated, including a large expansion and a smaller one. In this situation, a growth model is an interesting option. Parties may choose the smaller Maasvlakte, while keeping the option open—spatially, financially as well as administratively—to develop this smaller variant into the larger one at a later stage.

#### 4.4.5.6 Removing the Cause of the Conflict by Addressing the Underlying Question

The process architect examines whether there are any underlying factors that cause the persistence of the conflict. If he finds such factors, he designs an investigation or advice process that is to result in broadly supported conclusions about the validity and relevance of these factors.

In a discussion about the question whether reusable packages or disposable packages are better for the environment, there was indeed an underlying factor that dominated the discussion. This factor turned out to be the number of times a reusable package is used (‘the number of trips’). The parties were sharply divided about the environmental impact of reusable versus disposable packages. Some believed that the average reusable package goes through a large number of trips before it becomes useless. Reusable packages would thus be good for the environment. Others did not agree; they argued that the number of trips of reusable packages is low. The process architect who detects such an underlying factor may design a research protocol to solve this dilemma. The outcome surprised all parties: the sensitivity to the number of trips is highly limited. In other words, the number of trips appeared to be less relevant to the difference in environmental impact than the actors believed at the start of the process. This removed the cause of the conflict.

#### 4.4.5.7 Designing Mitigating and/or Compensating Measures

If the effects of particular options are inevitable and parties largely agree about their negative consequences for other parties, the process architect may outline a process that results in measures that mitigate or compensate these negative effects.

A process design is the result of a negotiation (Sect. 4.2). Cumbersome though this may seem, the above description of activities shows that there is not always a sharp distinction between designing and managing processes. Parties that reframe views as dilemmas and that negotiate about how these dilemmas should be addressed are not only designing a process, but have already started some of the substantive negotiations.

### *4.4.6 Process Dilemmas and Establishment of the Rules of the Game*

The process architect solves as many of the process dilemmas as possible in order to establish process rules that do justice to the two opposing sides of the dilemma. Experience shows that this is feasible for almost all process dilemmas. Examples of common process dilemmas include:

#### 4.4.6.1 Accuracy Versus Speed

Is the main concern for the process to be fast, with the associated risk of some degree of inaccuracy, or should it be designed in such a way as to be highly accurate, even if this is very time-consuming?

> The process arrangement that unites both extremes may be the following: 'Actors shall perform a quick scan to compare the alternative options available. If there is any doubt about the validity of the results of this quick scan, the parties may request a full, detailed analysis.'

#### 4.4.6.2 Many or Few Parties

Should there be many actors involved in the process or should the number of actors remain limited?

> The participating parties may all be given the same roles within the process. Alternatively, there may be a differentiation: there may for instance be parties that participate in the decision making, and parties that are heard or informed before any decisions are taken. The corresponding process arrangement may be: 'The process shall have an inner circle and an outer circle of actors. Certain issues shall only be addressed in the inner circle. The actors in the outer circle shall be asked to advise on the matter. If one of the actors in the outer circle reports a major interest in a decision, it shall be regarded as a member of the inner circle with regard to that particular decision.'

#### 4.4.6.3 Confidential Versus Public

Should research findings and the substance of advice that are generated in the course of the process remain confidential or will they be made public?

> The process arrangement may stipulate that everything is public, unless one of the participants makes an acceptable case for keeping certain elements confidential.

Once the problem and the dilemmas have been formulated, most of the sensitive issues have been described. The next step is that the rules are negotiated. Agreements will have to be made about the following issues:

- *Entry and exit rules*. The entry and exit rules describe which parties will participate, under which conditions parties are allowed to join the process, and how they can leave the process.
- *Decision-making rules*. These rules indicate how decisions are made: by consensus, for instance, or by a majority vote, in which case rules may be established about how to deal with the remaining minority votes. The decision-making rules tend to include agreements about conflict resolution. Conflicts may be settled by voting, by passing them over to another body, or by arbitration, to name but a few.
- *Organic rules*. These are rules that lay down the organization of the process. The process architect describes the bodies required, such as a steering committee, working groups and a group that monitors quality and progress. In some cases, these groups may require some kind of standing rules. Agreements also have to be made about the chairmanship and the secretariat of these groups. Last but not least, agreements are needed about the role of the process manager: what is his profile and which role does he play in the decision-making process?
- *Rules about planning and budget*. There is a plan that describes which activities will be performed in the process, in which order and with which deadlines. This plan also provides an estimate of the cost of the activities and the process management, and it describes who will bear which costs.

The concrete formulation of these rules will be guided by the design principles described in Chap. 3.

## *4.4.7 Testing the Process Design*

If needed, it is possible to test a process design. Again, the process architect may involve the parties in this, in order for them to become further committed to the process during the test.

Testing can of course be done in a variety of ways. The simplest way is a *prima facie* test, while the most elaborate way is simulation of the process that will take place.

### 4.4.7.1 Prima Facie Test

The architect organizes a brainstorming session about the rules that have been established. The future participants to the process examine the process design critically, focusing on whether it serves their interests and is likely to be successful. Does the process match the incentives and disincentives established earlier for each actor? Do actors have options for actions outside of the process that may be interesting for them? If so, which are the guarantees that the actors will remain committed to the process? Is there sufficient pressure on the process? Is the sense of urgency sufficient to commit actors to the process?

#### 4.4.7.2 Simulation

More detailed test results will become available through simulation. The architect develops a simulation that mimics the reality of the decision making as accurately as possible. The players in the simulation start out by addressing the process rules. Watching their behaviour, the architect analyses the functioning of the process rules and evaluates the results. Based on the results, he may adapt rules where needed.

A simulation may seem like an enormous investment, but there are two aspects that put this in perspective. Firstly, the simulation partly constitutes the real negotiation: experiences gained during the simulation may play an important role in the real negotiations. Secondly, if the parties wish to start out by going through a simulation, the process architect is not in a position to ignore this wish. Parties may feel the need for a simulation particularly when there is much at stake, or when there is a large degree of distrust.

### *4.4.8 Participation*

The final activity is gathering the participants to the process. This is of course a crucial step: after all, the process largely depends on the behaviour of individuals. Several aspects play a role with regard to participation.

#### 4.4.8.1 Involved in Designing the Process, or Not?

As we observed earlier, designing a process is an important learning process for all parties. That is why it may be desirable that the people who will participate in the designed process, have been involved in it. Alternatively, it may be desirable to appoint other people as representatives in the process, for instance if the process design was characterized by much conflict.

#### 4.4.8.2 Right of Consent with Respective Appointments?

This right implies that a person may only join a process on behalf of a party if the other parties agree with this. One advantage of this may be that there is less risk of incompatible characters, and also of asymmetrical representations.

> The organizations that participate in the process may have different interests in the process. Some parties are passionate: their interests in a successful outcome of the process are significant; other parties are less passionate. This may have important implications for the choices parties make with respect to their representatives. Chances are that the passionate party will send a heavyweight representative, for instance from the organization's sub-top, while the other party sends an intermediate-level representative. Such asymmetry may disturb the process.

A disadvantage of this method is of course that it enables parties to interfere with the internal affairs of other parties or make strategic use of this arrangement.

#### 4.4.8.3 Direct or Indirect Representation?

Direct representation means that a party appoints a representative. Indirect representation means that a person participates in a process on behalf of a party, without acting as its formal representative. This indirect type of representation is used by parties that feel it important that a particular party should participate in the process, while this party cannot be persuaded to do so. This indirect representative will then have to 'earn' his 'own' party's commitment during the process.

For each of these dilemmas, there is no choice that is a priori right or wrong. A choice is right if the parties conclude during the process that the choice is right.

## References

1. Curran D, Sebenius JK, Watkins M (2004) Two paths to peace: contrasting George Mitchell in Northern Ireland with Richard Holbrooke in Bosnia-Herzegovina. Negot J 20(4):513–537
2. De Bruijn JA, Geut L, Kort MB et al (1999) Procesvoering besluitvorming nationale luchthaven, Den Haag, commissioned by the Dutch Ministry of Transport, Public Works and Water Management
3. De Bruijn JA, ten Heuvelhof EF (1995) Ontwerp besluitvormingstraject Maasvlakte II, Delft, commissioned by GHR
4. De Bruijn JA, ten Heuvelhof EF (1996) De implementatie van het Verdrag van Malta: Een ontwerp voor procesmatige sturing, Delft, commissioned by the Dutch Ministry of Education, Culture and Science
5. De Bruijn JA, ten Heuvelhof EF (1997) Ontwerp van een protocol voor de toetsing van ALARA-inspanningen, Delft, commissioned by Foundation Merk Artikel
6. De Bruijn JA, ten Heuvelhof EF (1999) Procesbenadering voor productgericht milieubeleid, Delft, commissioned by VNO-NCW
7. De Bruijn JA, ten Heuvelhof EF (1999) Kwaliteitssysteem beleid en beleidsprocessen, Delft, commissioned by the Dutch Ministry of the Interior and Kingdom Relations
8. De Bruijn JA de, ten Heuvelhof EF (1999) Package covenant II: clustering and process of reporting protocol. Report for the Ministery of Housing, Environment and Spatial Planning and the Dutch Packaging Industry, Delft, commissioned by Foundation Verpakking en Milieu
9. De Bruijn JA, ten Heuvelhof EF (2000) Procesontwerp onderzoek reisgedrag: NS reizigers, streekvervoer en VSV+, Delft, commissioned by the Dutch Ministry of Education, Culture and Science
10. De Bruijn JA, ten Heuvelhof EF (2000) Procesafspraken voor de onderhandelingen tussen het ministerie van Onderwijs, Cultuur & Wetenschappen en de openbaar vervoerbedrijven over een nieuwe reisvoorziening voor studenten, Delft, commissioned by the Dutch Ministry of Education, Culture and Science
11. De Bruijn JA, ten Heuvelhof EF, de Vlaam HIM (1999) Interconnection disputes and the role of the government between substances and process. Commun Strateg 34:295–317
12. De Bruijn JA, ten Heuvelhof EF, van Eeten MJG (1997) Quick scan van procesontwerpen voor vervlechting van dialoog, onderzoek en besluitvorming rondom TNLI, Delft, commissioned by TNLI

13. De Bruijn JA, in ’t Veld RJ et al (1999) Procesvoering projectdirectie Mainport Rotterdam, The Hague, commissioned by project executive board Mainport Rotterdam
14. Fisher R, Ury W (1981) Getting to yes. Houghton Mifflin, Boston
15. Kotter JP (1995) Leading change: why transformation efforts fail. Harv Manage Rev 73(2):59–67
16. Termeer CJAM (1993) Dynamiek en inertie rondom mestbeleid. Vuga Uitgeverij, Den Haag
17. Van Twist MJW, Edelenbos J, van der Broek M (1998) In dilemma’s durven denken. Management en Organisatie 52(5):7–23

# Part III
# Managing the Process

# Chapter 5
# An Open Process

## 5.1 Introduction

This chapter examines ways for the process manager to ensure that the decision making is an open process: the relevant parties have to be involved in the decision making and they must be certain that their interests will be addressed where possible, in accordance with the process agreements. This implies that the initiator as well as these parties should be involved in drawing up the agenda of the process.

We will therefore take a closer look at the three design principles relating to the openness of the decision making, which were introduced in Chap. 3.

Section 5.2 will address the design principle that is relevant at the start of the process approach: involving other parties in the decision making. A number of criteria that can be used here have been mentioned in Chap. 3. In this chapter, some follow-up questions will be addressed. There is a certain tension between the need for openness and the need for control. When too many parties are involved in a process, the result may be utter indecision: much discussion, little progress. The process becomes completely uncontrollable. How to deal with this?

The next question is: what can be done when an initiator is not sure which parties are relevant, since this only becomes clear during the process? And: what can be done if an initiator needs these parties, but they are unwilling to participate in a process?

Section 5.3 examines the transformation from substance to process—the second design principle. When can a process manager apply this principle, and how should he avoid the risk of a process falling victim to proceduralism? We will also address the question whether a process benefits from a substantive framework.

Section 5.4 deals with the third design principle: the requirement that the process should be transparent. We will focus particularly on the various potential roles of a process manager.

H. de Bruijn et al., *Process Management*, DOI: 10.1007/978-3-642-13941-3_5,

## 5.2 Involving Parties in the Decision Making

In Chap. 4 we have already discussed the criteria for involving parties in the decision making. This section will address three issues in the context of this design principle: controlling a process; the unrecognizability of actors, interests and resources; and parties' potential refusal to participate.

### *5.2.1 Controlling a Process*

#### 5.2.1.1 Openness of the Agenda

The principle of openness implies that the parties that are to be involved in the process are also involved in drawing up its agenda. After all, they will only join the process if it is interesting for them; the process becomes interesting when subjects are discussed that touch upon the parties' interests. (The mechanism behind this is referred to as 'Enlarge pie first, cut later'; see for instance Susskind [13] and Innes and Booher [9]).

As a result, the agenda of the process may be more comprehensive than the initiator originally intended—it will be a multi-issue agenda. Moreover, new agenda items may sometimes be a reason to invite new parties because of their production power regarding these new issues—which will cause the process to expand even further.

Initiators tend to react to this in an intuitive way, arguing that the process becomes too complex, that it is less likely to produce results, and that it is therefore imperative to reduce the agenda to a limited number of important items as soon as possible.

From a project perspective, this is an attractive strategy: it makes it easier to control the process. From a process perspective, however, it is preferable to maintain the multi-issue nature of the decision making as long as possible. There are three reasons for this.

Firstly, an agenda containing a large number of items allows for coupling of these items. Party A has an issue that can be solved by party B because it has the required resources to do so. Party B is prepared to make these resources available, because party C will then no longer obstruct an option that is appealing to party B, and so on. This strategy has proven to be effective in cases such as the GATT negotiations, "where concessions in one industry could be traded against those in others" [2, p. 556].

Secondly, a multi-issue agenda nearly always reduces conflict. There is a simple explanation for this: for every issue there will be different coalitions of supporters and opponents. If this is the case, it will be hardly appealing for a party to stir up a conflict about a particular issue; after all, this other party may be a coalition partner in another issue. This calls for moderate behaviour towards this partner.

In the earlier-mentioned programme 'More space for rivers', there is a moment when the State Secretary does not receive much support for her project proposals yet. This is because these proposals evoke quite some *Not In My Back Yard* (*NIMBY*) reactions: local and regional authorities support the policy in general, but they keep objecting against concrete projects that are to be carried out in their jurisdiction. One of the projects proposed by the national government is the relocation of a business area that is currently located in a town in the riverbed, on the land side of the dike. Removing this business area will benefit the flow of the river water. The mayor, who has no objections *per se* against the project 'More space for rivers' in general, opposes this project, and therefore it looks like its implementation will be a difficult, sluggish process.

At a certain moment, the State Secretary declares that she would not object if the local authorities decided to couple the 'More space for rivers' project to their own, local goals. The mayor accepts this idea. During some time, he has been worried about the accessibility of the business area to the latest generation of cargo ships. The substantial draft of this kind of ships prevents them from reaching this business area, which reduces its value. In response to the State Secretary's announcement, the mayor commissions the development of an alternative plan for the business area. In this new plan, the business area will continue to exist, but it will be surrounded by a system of deep sludges that will both allow more space for the river, and make the business area better accessible to ships with large drafts. The State Secretary approves this plan. The result is a win–win situation, with the mayor becoming a supporter rather than an opponent of the project. In conclusion, while adding issues to the agenda may make the projects substantively more complex, it also makes it easier to gather managerial support. This facilitates the implementation.

Thirdly, it is important that the BATNAs [6] of the relevant parties are taken into account during negotiations. A party's BATNA is its Best Alternative to a Negotiated Agreement. If a party has an attractive BATNA, it will hardly be willing to settle or make concessions. Of course this will not be conducive to the progress of the negotiations. Adding issues to the agenda may shift the balance of BATNAs. After all, an actor may have an attractive BATNA in relation to issue A, while its BATNA regarding issue B is a lot worse. Combining A and B in the negotiations will make this actor more willing to compromise with regard to issue A, since this may help him to avoid his BATNA when it comes to issue B [15, p. 28].

#### 5.2.1.2 The Tension Between Openness and the Need for Control

Although complexity (in the sense of many parties putting forward many items for the agenda) may promote decision making, as we have seen, there is of course a limit to the number of parties and issues to be addressed during the process. In other words, there is tension between the openness of a process on the one hand and its controllability on the other hand. How to deal with this tension? Two types of strategies are possible.

The first strategy is that prior to the process, the initiator selects the parties to be involved and also forms an opinion about the selection of items to be placed on the agenda. The advantage of this procedure is that the process designed is controllable, but a major disadvantage is that the excluded parties may oppose the process. Moreover, it is often not easy to make an a priori estimate of the parties that are important and those that may be excluded. Something similar is true for the agenda of the process. An issue that may seem like a detail at the start of a process, and that therefore does not need to be placed on the agenda, may become a main issue during the process. In this case, "toxic assets" [14, p. 153] are of particular importance. These are issues that are so contentious that they are bound to block progress regarding other issues. These assets should therefore always be avoided during the process. This is why this first strategy is effective only if the initiator can be absolutely sure that the limitations he imposes will not be subject to any discussion during the process.

A second strategy is that at the start of the process, the initiator does not yet assign any role for the value of control. He generously invites parties and also generously allows them to put forward items for the agenda. The underlying idea is that control should not be imposed on the parties beforehand, but that it will develop *during the process*. The parties will learn during the negotiations that the open nature of the process makes it insufficiently controllable. They may conclude, for instance, that the large number of items on the agenda hampers decision making. In such a situation there are two possibilities:

- A critical mass of parties has no interest in taking decisions and is satisfied with the stalemates that arise.
- A critical mass of parties is dissatisfied with the stalemates that arise and wishes to make arrangements for enhancing the controllability of the process.

In the first situation, there is apparently no support for taking decisions. The process architect has selected parties and agenda items and has done so in consultation with the initiator. Apparently the initiator has also concluded that the selected parties and items are necessary to serve his interests. The conclusion should then be that there does not appear to be sufficient support yet to arrive at decisions quickly.

In the second situation, support will arise during the process for proposals to improve the controllability of the process [14, p. 157 e.v.]:

These proposals may for instance imply that certain parties settle for a marginal role in the process, or have this marginal role imposed on them by other parties. It may be an interesting option to let some parties be represented by other parties, as will be explained below. It is also an option to temporarily remove certain items from the agenda, to rephrase them, or rather to deal with those quickly. In the context of international diplomacy, in such cases 'shuttle diplomacy' may be preferable to a 'summit'. In the latter case, the agenda tends to be overburdened, all items are coupled to others, and an obstruction regarding one of the issues may block the entire summit, including agreement on easier issues [14, p. 157 e.v.].

Finally, there is the possibility to distinguish different roles in the process: taking decisions, listening, providing advise, and so on. The point is that such forms of control are more likely to be accepted by parties, since they evolved during the process and were suggested by the parties themselves (at least, by a critical mass of parties). In the hypothetical case that an excellent process manager comes up with exactly these same control measures at the start of the process, arrangements made during the process still remain superior to preplanned arrangements. After all, something that originated during the process will enjoy the parties' support, while something that was preplanned will not. As soon as a preplanned arrangement gives rise to any disadvantages, the parties will blame the process manager for this and he will be expected to solve the problem. If an arrangement made in the course of the process creates any disadvantages, on the other hand, the parties will have to solve this problem together. Moreover, there will always be parties that dislike the role that is assigned to them. A division of roles will have more authority, however, if is established by the entire body of parties rather than by the process manager or—even worse—by the initiator. In the first case, parties that oppose their assigned role will have conflicts with all other parties, while in the latter case these conflicts will only involve the process manager or the initiator.

### 5.2.2 The Unrecognizability of Actors, Interests and Resources

So far we have assumed that the process architect and the process manager are able to get an impression of the dependencies in a network and of the actors that are to be involved. It may, however, be unclear which actors need to be involved in the decision making, and which interests and resources they have. It may also be unclear which actors are necessary to enrich the decision making.

There is another dimension to the unclarity of the interests: it may become apparent *during* the process that the process *architect* has made an incorrect estimate of the relevant actors, their interests and their resources. Actors that ought to play a role because of their interests and/or power may not have been admitted to the process.

This may jeopardize the process. Adjusting the process (involving new actors, making new agreements) may add to the chaos. Failing to adjust the process, however, may cause new actors to feel insufficiently represented in the process—which may sometimes surprise the process architect.

An example to illustrate this is the earlier-mentioned negotiation process between business and civil society with regard to packages. Some time after the start of the process, business requests to appoint an extra representative in the steering group, on behalf of the non-food packaging/filling industry, given the fact that the seat for packaging/filling is taken by the food sector.

The underlying idea is that the interests of the non-food sector cannot be sufficiently taken into account by a representative from the food sector. The request is denied, because it would disturb the balance in the steering group.

In principle, this is a risky development: an important party finds itself underrepresented in the process. This may affect the open nature of the decision making. However, the refusal to include the non-food representative did not in fact jeopardize the process. An important explanation for this is that the business representative had organized consultations with the group he represented very systematically. Before each meeting of the steering group, consultations were held with representatives at the sector or company level in a 'round table group'. This group was accessible to all relevant companies. This gave the companies a permanent account in the process.

Something similar was the case for the representatives of the Consumers Association and the environmental organizations. Although the groups they represented were not organized, they did constitute the account for consumers and environmental interests, respectively [3].

The problem of the unrecognizability of the main actors, their interests and their resources may thus be solved by:

- having the process architect formulate the relevant interests at a high level of abstraction ('environment', 'industry', 'consumers');
- giving the representatives in the process the role of account: any views held within the wide interest they represent can be put forward through them;
- giving the representatives in the process no detailed mandate beforehand; this may be a counterintuitive arrangement, because it seems to evoke non-committal attitudes, but the advantage is that if there is any change in the actors' involvement, representatives have some room to maneuver;
- ensuring a heavy representation, allowing the representatives to play an authoritative role also towards those they represent and to convince them of certain standpoints (they are not only the 'messengers' of those represented, but can also play the role of 'missionaries' towards them);
- giving the representatives in the process the possibility to 'earn' the mandate and the commitment of those represented during the process—for instance by allowing them 'quick wins'.

Table 5.1 lists the differences between an 'account' and a classical representative role.

The same mechanism that applies to the question of control is also present here. Control evolves during a process; in much the same way, a mandate may evolve. If a mandate is to be arranged beforehand and in detail, this may cause much hustle

**Table 5.1** Differences between a representative and an account

| Representative | Account |
|---|---|
| At party level | At interest level |
| Much commitment power | Little commitment power |
| Clear mandate | No mandate or an unclear mandate |
| Staffing: usually of a lower standing than the represented: the 'messenger' | Heavy staffing; able to play the role of 'missionary' towards the represented |
| Commitment by those represented is secured prior to the process | Commitment by those represented results from the process, has to be earned |
| Is able to cope with dynamics if and insofar as the represented permit this and/or it is reconcilable with the mandate; may be problematic | Is able to cope with dynamics if and insofar as the broad description of the interest permits this; this will nearly always be the case |
| Given the strict mandate, expansion of the number of those represented (and thus of those who determine the use of the mandate) will be a problem | Given the loose mandate, expansion of the number of those represented (and thus of those who determine the use of the mandate) will be no problem |

and bustle. Allowing a mandate and a level of commitment to evolve during the process, however, reduces the amount of trouble beforehand, and allows the substance of the mandate to move along with the process.

### *5.2.3 Parties' Refusal to Participate*

There may be a sharp difference between participation at the start and at the end of a process. At the start of the process, some actors may have a limited level of interest. One reason not to participate is a favourable BATNA, which hardly provides a positive incentive to participate while the costs may be considerable. A process manager should therefore invest in an analysis of the BATNAs of desired participants. A second reason may be that these actors are insufficiently aware of the direction the decision-making process will take, while participation will cost them time, money and perhaps concessions. This means that there is insufficient incentive for some actors to participate, even though at this initial stage there is plenty of opportunity for them to influence the decision making.

Towards the end of the decision-making process, the situation is exactly the opposite. Some actors will be very interested in participating in the decision-making process at this stage; after all, the results of the process are gradually becoming clearer. At the same time, however, the main decisions have already been taken, which limits the possibilities to exert influence. In other words, at the moment when there are ample opportunities to influence the decision making (i.e. at the start), the degree of participation is low, while it is high at the moment when the opportunities for influencing are limited.

> Drawing up a traffic plan for a certain area requires the involvement of many actors: local authorities, the province, companies and citizens. At the start of the process it is often unclear what the eventual traffic plan will imply. This vagueness is a reason for particular actors not to participate in the decision making, or to participate only formally. Once the plan is shaped in more detail, certain actors notice that their interests are insufficiently addressed, and they start interfering with the decision making at this late stage. A consensus between the participants that was perhaps reached after laborious negotiations is thus reopened for debate.

Actor behaviour as described above—low degree of participation at the start, high degree of participation towards the end of the process—is of course a threat to the decision-making process. Ideally, a decision-making process evolves from variety (start) towards selection (end), as has been shown in Sect. 3.6 in Chap. 3. If participation is low at the start, this may mean that insufficient variety is being generated. Towards the end of the process, authoritative selection is made impossible by actors suddenly wishing to participate, thus generating new variety.

It is important for the process manager to prevent this. Firstly, this mechanism confirms the need of a *sense of urgency*; a process that is started too early tends to be ineffective.

Secondly, this mechanism confirms the need for clear process agreements. After all, the lower the number of explicitly defined process agreements, the higher the chance that this mechanism presents itself.

Finally, it is important for the process itself to do the work wherever possible. The process involves parties and those parties have networks. Parties may use their networks to involve unwilling parties in the process. Perhaps informally at first, through a modest amount of information and informal consultation, but with a little more pressure later on. After all, a critical mass of parties reaching agreement in a process is often quite powerful. An unwilling party may be approached from several sides at the same time, and may thus be convinced to participate.

## 5.3 The Transformation from Substance to Process

One of the main design principles stipulates that the number of substantive choices being made prior to the process should be limited as much as possible. All that is laid down is the *moments of choice* in the process, and how the decision-making *process* will pass off at these moments. The idea is that the parties in the process make substantive decisions at the moments of choice defined beforehand, in accordance with the previously agreed process agreements.

This design principle may also play a role *during* the process. An intended substantive decision may be transformed into a process-related decision. This mechanism will be addressed in this section. We will describe under which circumstances such a transformation takes place, what the risks are, how the process manager may deal with these risks, and whether and how the process manager should draw up substantive frameworks.

### *5.3.1 Conditions for a Transformation from Substance to Process*

The transformation from substance to process implies that parties should agree on a procedure that is to be followed when a substantive choice is to be made. For example, parties that have to choose between A and B decide that they will start by examining possible alternatives, then add these alternatives to A and B, and then make a decision. Such a transformation is necessary if parties are unable to make a substantive decision together. This will be the case in the following situations:

- when there is serious distrust between the parties;
- when the parties are insecure about—or not yet accustomed to—the process;
- when there are many substantive uncertainties among the parties;
- when conflicting interests cannot be reconciled (yet).

In the earlier-mentioned negotiation process between business and civil society about packages, this mechanism is strongly present. In the first few months, virtually every substantive decision is avoided, and rather transformed into a process decision. The representatives of these organizations meet for the first time in the context of the process. They have not built up a substantive frame of reference yet, nor grown accustomed to the decision-making process. It will be evident that they are hesitant to make substantive decisions for that reason [3].

There is an important risk associated with continuously transforming substantive decisions into process decisions. Parties may get the idea that no progress is made in the decision making, and that the process is falling victim to proceduralism. Criticisms such as 'too little substance', 'indecision', 'sluggishness' and 'too much talk' can be fatal to a process approach.

This may make the situation very difficult for a process manager. After all, insecurity and distrust in a network make it impossible to make substantive decisions. What's more, due to unceasing procedural decisions all parties still have prospects of gain during the rest of the process. This promotes their trust in the process and also offers them an opportunity to build up trust in each other.

When parties aim for substantive decision making anyway, however, chances are that the process will drown in an abundance of details, and thus fail at an early stage. After all, insecurities and distrust (for lack of a shared frame of reference) quickly result in parties assuming an attack pose: every decision has to be secured. Every party will fear that a particular decision will limit its degrees of freedom or create a *point of no return*.

The process manager therefore has to navigate between the Scylla of being reproached for proceduralism and the Charybdis of premature substantive decision making. In other words, he should ensure sufficient support for the process approach among the participants, even though it is impossible to make substantive decisions. Parties should learn how to recognize the advantages of procedural decisions. The following strategies may be helpful in this regard.

- *Every party benefits from the process agreements; the process manager makes the benefits of the process agreements visible to everyone.* In the first place, internalization can be achieved by making clear to the participants that they each have an interest in—or will benefit from—the design principle 'substance becomes process'. Chances are that party A will advocate substantive decision making if it feels (1) that a substantive decision serves its interests at that moment and (2) that it might find a majority in favour of a decision. Such a decision may come too early in the process because it may harm the interests of the other parties too early in the process. At such a moment, party A will hardly support making procedural decisions in accordance with the design principle 'substance becomes process'. Party A will have to conclude at some stage that other parties do find themselves in this comfortable position, which will jeopardize party A's position. The process manager may have to point out to the participants that all of them may benefit from the design principle. This may create support for the process design as well as some reserve towards premature substantive decision making.

- *Tolerance of irrelevant, but substantive decision making.* An important mechanism in processes of this kind is that substantively interested parties have a limit of tolerance above which they no longer accept process management unless it is accompanied by substantive discussion and decision making. If this limit is exceeded, there is a serious risk that the process approach will be blamed for the shortcomings mentioned above. Consequently, support can only be retained by ensuring a regular dose of substance: a substantive discussion followed by decision making. In a particular situation, it may be clear to the process manager that such decision making will be revoked in the course of the process. For the sake of support for the process, it may be useful to show some degree of tolerance towards substantive discussions that are not directly related to the final process result. In such a situation, too, a process manager should not wish to exert too much control. Rather, he should allow the process to take its course.

### *5.3.2 Ambiguity*

The tension between substance and process is clear. From the process perspective, it is useful to minimize the number of controversial substantive agreements at the start of the process. The process manager should transform substantive issues into process agreements whenever possible. However, there is a limit to this. At some point, there is too much 'process', and the need for substantive agreements and substantive progress will be so substantial that it becomes impossible to make further process agreements. One way to resolve this tension between substance and process is to make agreements that are substantively ambiguous. Such agreements are substantive by form, but given their ambiguity they will have to be elaborated at a later stage, which will call for a process after all. Substantively ambiguous agreements include concepts such as 'quality', 'balanced', 'sustainable', 'gradual' and so on. The use of such concepts in agreements may set afloat processes that have become stranded. The 'feel good' nature of such concepts easily evokes parties' approval—although they may each assign their own meaning to this. These interpretations may of course lead to confrontations at a later stage, but at that moment the introduction of another 'Sunday concept' is likely to advance the process once more. This 'constructive ambiguity' may be appealing to the individual parties, because it puts them in a position that allows them to present the agreement to their supporters as a victory [5, p. 193].

A cynic may note that such agreements have little value, since all they do is grant a delay. However, such cynicism ignores the dynamics of these processes. After all, even an ambiguous agreement may reinstall the momentum and élan in a process. There may be room again for trust in each other and in the process, which allows for future agreements that were initially assumed to be impossible. Another possibility is that, following the ambiguous agreement, the process takes an entirely different course, with the ambiguity never returning to the negotiation table. In that case the ambiguous agreement has fulfilled its positive role, while the risks remain unnoticed.

### 5.3.3 Frameworks and Crystallization Points

A proper process offers sufficient room to the relevant parties to define problems and find solutions together. An important question in this regard is whether and how substantive preconditions can be imposed on such a process. This involves a substantial dilemma.

On the one hand, processes only stand a chance of success if parties are offered an opportunity to define problems and solutions jointly. This is risky for certain parties: they clearly prefer certain problem definitions and solutions, and they are not sure whether these will survive the process. These parties will have to participate in a process all the same, because this is the only way for them to earn the support of other parties.

One attractive way for parties to deal with this tension is by formulating preconditions beforehand. This limits the degrees of freedom of the other parties and thus increases the chance of an outcome that is sufficiently appealing to them. If the process approach is used to accomplish a certain change, certain parties thus have a strong incentive to demand predefined preconditions.

The other side of the coin will be clear immediately. Predefined preconditions may deter the other parties. These may get the impression that the process is framed in preconditions in such a way that there is insufficient room for their own interests to be served, or for creative, unforeseen solutions to emerge. The process becomes oppressive—parties experience it like a funnel trap: in the end there is only one possible direction for the process, and there is no way back.

This triggers the question how a process can be subjected to substantive preconditions without becoming oppressive. In that regard we will introduce a distinction between frameworks and crystallization points. This question will be further addressed in Chap. 6.

A *framework* offers room within predefined limits. Those who recognize that ex ante substantive planning of an activity is impossible should offer room to others. However, this room cannot be unlimited. Therefore this room is quantified ex ante through the formulation of frameworks. In other words, there is room for innovation and creativity, but within predefined frameworks. Frameworks, however, can all too easily develop into rigid preconditions, which may be perceived as impediments and which may frustrate innovation (Table 5.2).

*A crystallization point* is a substantive idea that can be further elaborated. A process that is designed around a number of crystallization points invites parties to discuss these crystallization points, to criticize and enrich them, and to couple

**Table 5.2** Frameworks and crystallization points

| Frameworks | Crystallization points |
|---|---|
| Development *within* … | Development *from* … |
| Substance constrains | Substance challenges |
| Incentives for the absence of the party that establishes frameworks | Incentives for the presence of the party that establishes frameworks |

them to other crystallization points. Crystallization points differ from frameworks in three respects.

The point of a framework is to ensure that a development stays *within* certain substantive limits. Crystallization points, on the other hand, are aimed at the *development of* certain substantive notions. A framework is binding and sets limits: the process has to proceed within the framework. Crystallization points offer possibilities: in the process, the crystallization points are further developed in a direction determined by the parties.

A party that has a highly substantive orientation and that translates this orientation into frameworks will have an oppressive influence on the other parties. If this same party translates its substantive orientation into crystallization points, on the other hand, it will stimulate other parties. The richer the substance of a crystallization point, the stronger the stimulus it provides to the other parties.

The third difference—which is not intrinsically present—is that a party that establishes frameworks is not necessarily active in the process. After all, the frameworks offer sufficient guarantees that the outcome will stay within the bandwidth acceptable to this party. For a party that formulates crystallization points, however, there is a strong incentive to participate in the process. After all, it will wish to be involved in the further elaboration of these points.

A family that has to make a decision about its holiday destination has four different options.

- The first strategy is substantive: the parents decide that the holiday will be spent in Sardinia from 1 to 21 August and that the family will travel by air.
- The second strategy is a process; the parents indicate that a decision about how to spend the time between 1 and 21 August will be taken in a consultation process in accordance with certain rules.
- The third strategy is also a process, but the parents establish frameworks: the holiday has to be spent in Italy, the family will travel by air and destinations that the family has visited before will be excluded.
- The fourth strategy also involves a process, but it is shaped around crystallization points, such as an island, sunshine and culture. An interaction process will develop, starting from these three crystallization points, in which the family members consider a number of options in mutual consultation. This interaction process may take a surprising turn. It may end with the conclusion that the holiday will be spent in Northern Norway. One family member proposes Sicily, because of the rugged mountains in the north of the island. In the following interaction process, the family conclude that this is an appealing option: mountains close by the sea. However, the intense sunshine in Europe's far south would be a problem for another member of the family; thus they

> continue looking for a combination of coast and mountains. Now Northern Norway suddenly becomes an option—the days are long, the temperature is often pleasant, and the mountains are close to the sea, which creates an island feeling. Had 'island', 'sun' and 'culture' been preconditions, Norway would never have been chosen. Now that these are crystallization points, however, this option is indeed possible. At the same time, the crystallization points do offer the process initiator some support—perhaps not as much as preconditions, but in any case more than in a completely open process.

Of course, the fourth approach does not imply that anything is possible. There will be budget constraints on the choice of a holiday destination, for instance. In the first approach, the parents have already taken this into account. They can determine a substantial part of the cost of the holiday by choosing the destination and the mode of travel unilaterally. In the third approach, they can lay down the budgetary preconditions in the framework: the holiday should not cost more than an x amount of money. This is different in the second and fourth approaches. During the consultation process, the parents will indicate that they, as the financers of the holiday, have an x amount available. Since they provide the funds, a precondition *develops during the process*, but it need not be a rigid one. After all, the family members may eventually decide that they will skip this year's winter vacation and that the budget thus made available will be spent on the summer holiday, allowing the choice of another destination.

This indicates, by the way, that in the absence of frameworks there is a strong incentive for the initiating party—in this case the parents—to participate in the process. Without their participation, the cost of the holiday might easily get out of hand.

This example illustrates a complicated matter. What if the process initiator cannot live with the outcome? In this example, the parents are the initiators. What if Norway is absolutely unacceptable to them as a destination? The first answer: they participate in the process and they make their objections clear to the other participants—if needed they can even block the decision to go to Norway. But this may be insufficient. Particularly if there are several—and other powerful—parties that participate in the process, things may take a turn that is entirely different from what the initiators expected. How to deal with this? Which measures to take? There are several potential measures that can make the process a safe one for the initiator too—as will be explained in Chap. 6.

## 5.4 Process and Process Management are Characterized by Transparency and Openness

An important building block for the process approach is the notion of transparency: a transparent and fair process design makes entices parties to participate in the process. Transparency means that they can check whether the process is fair and whether it offers them sufficient opportunities to promote their interests.

Transparency may also be required with regard to the role of the process manager. An important notion in theoretical discussions about resources is that the process manager plays an independent, disinterested role in the decision-making process, and derives much of his authority from this principle [1, 10].

### *5.4.1 The Roles of the Process Manager: Dependent and Independent*

The idea that a process manager may have interests seems to include a contradiction: after all, the process manager is the disinterested facilitator in the process. He holds an independent position towards the other parties.

In reality, the relationship between the process manager and the other parties is more paradoxical. The process manager is independent, but also dependent on the parties in the process. If the process manager is to perform his task, he will need the parties' support. The result is a very subtle relationship between the process manager and the parties. He is supposed to act transparently and to assume an independent position among the parties, but at the same time his functioning is dependent upon these parties. When one or several parties give up their support for his authority, he is vulnerable. How can he deal with this sometimes fragile position? How can he develop sufficient authority as a process manager?

One strategy to deal with this is reinforcing the process manager's position of power by giving him additional functions. A number of such functions are suggested in the literature:

- making a substantive contribution: the process manager is also an expert on substance;
- managing financial resources: the process manager is also the treasurer; and
- keeping the balance: the process manager as 'countervailing power', as 'balancer', embodying particular underrepresented interests in the process.

These roles are attractive in a *cooperative* environment. The process manager can make a substantive contribution, he can exercise some financial steering and he can maintain the balance in the process in his role of 'balancer'. Such an accumulation of functions will also make the parties more dependent on the process manager, which may reduce the vulnerability of his position. In short, an accumulation of roles will enhance the problem-solving power of the process manager.

In a *non-cooperative* environment, however, these roles may turn against the process manager:

- substantive statements may be perceived as a choice in favour of one of the parties;
- the process manager may be blamed for any financial problems;
- a 'balancer' may be reproached for opportunism or partiality;
- the various roles may interfere in an unfortunate way: the position of a substance expert may differ from that of a 'balancer'.

If a process manager is given such roles anyway, some familiar displacing phenomena may occur: substance displaces process, money displaces process, power displaces process. Each of them poses a risk to the process manager's position, which results in a paradox: the more resources a process manager has available, the greater the risk that they will harm his position.

Rosabeth Moss Kanter has drawn up a list of the skills of a change manager ('change agent'), four of which we quote here [10]:

- 'the ability to perform effectively, without the power, sanction and support of the management hierarchy;
- the ability to develop high trust relationships…;
- respect for the process of change, as well as the content;
- the ability to work across business functions and units, to be "multifaced and ambidextrous".'

Remarkably, formal power is seen as a disadvantage, and substantive expertise and process expertise are regarded as equally important.

One may wonder how the process manager can protect his position in a non-cooperative environment. This is where a second strategy presents itself. In essence, this strategy implies that a process manager should rather adopt a reserved attitude.

He confines his role to one single function (managing the process) and thus renounces the other roles. This reduces his power to solve problems himself. As a result, he will have to leave these problems for the other parties to solve jointly. After all, they are the ones who will have to support the final decision; they can only do so whole-heartedly if they feel ownership over the solutions to their conflicts. There are several possible strategies to accomplish this.

#### 5.4.1.1 Naming and Framing

The process manager is in a position to summarize or describe the parties' viewpoints, dilemmas in the process or conflicts between parties in a certain way. This naming and framing of issues may affect the parties' behaviour—they are either stimulated to participate in the process, or they are not. Naming and framing may either stir up resistance, or not.

*Framing 1.* The need for framing is illustrated by an analysis of the negotiations between the US and China about Intellectual Property Rights (IPR) [8]. The challenge facing the US was so significant that it didn't seem to have a chance of success. In the first place, there was of course adamant Chinese opposition. But support from within the US was not unanimous either. Particularly non-business US interests put up resistance against a firm attitude towards China. These groups valued China's support against states such as North Korea and Iran. Human rights groups did not want to waste US negotiating capital on 'low-level commercial considerations'. And even some business sectors in the US did not favour a firm approach towards China. Boeing, for instance, regarded China as an interesting growth market, which it did not want to jeopardize by a firm position on IPR. The foreign partners of the US supported the US in principle, but repeatedly showed to be unwilling to really turn against China, afraid as they were to harm their own interests in China. The US policy, however, managed to become successful anyway thanks to a strategy of "framing the issues, carefully tailoring the frame for various audiences" [8]. Every target group received its tailor-made message. US human rights groups and business, for instance, were made aware of the fact that IPR constitute an example of rule of law, which would also serve other interests, including human rights and business interests.

Within China, too, several target groups were identified which each got their own message. Many Chinese people, for instance, were in favour of China's accession to the WTO. For this reason, the IPR negotiations were coupled to China's WTO membership. Much effort was also put into "enhancing the credibility and salience of sanctions" [8, p. 321].

*Framing 2.* Policy aimed at addressing the problem of energy shortage can easily be framed as a search for options to reduce energy use. This may sound self-evident, but for many parties this is not an appealing option. An alternative framing would be that energy shortage calls for 'ecological modernization' [7]. To business and the engineering world, this is a more interesting perspective. They are invited to think about new possibilities. From the perspective of business and engineering, this is much more appealing than economic shrinkage—and thus more likely to receive support.

#### 5.4.1.2 Using Events Outside of the Process as Policy Windows

Of course a process manager focuses primarily on the process that he is working on. There may be developments outside of the process, however, that he can use to stimulate the process—events that he has no direct influence on, but that he should be aware of as they could be beneficial to his process.

After six countries signed a treaty in 1951 establishing the European Coal and Steel Community, it appears to be difficult to broaden and deepen this cooperation—i.e., to involve additional countries in the Treaty, and to address other issues besides coal and steel, respectively.

Broadening is a problem particularly with regard to the UK. The UK does not support European integration. From the British perspective, the UK is at least as connected to the US as it is to Europe. The Suez crisis, however, changes this situation. In 1956, Egypt decides to nationalize the Suez Canal. France and the UK do not accept this, and invade Egypt. To the UK's great surprise, the US refuses to support its 'natural' ally. From this moment on, the UK understands that it should not place too much trust in the loyalty of the US. Instead, it develops an increasing interest in the developing EU, even though it will be another 17 years before the UK actually accedes. In other words, for the EU the Suez crisis represents an opportunity.

Deepening the cooperation turns out to be problematic as well. It is particularly Germany that has its doubts. The Russian invasion of Hungary, however, removes all doubt in Germany about its international future. Germany realizes that it is only Western Europe that offers an interesting future perspective. Chancellor Adenauer therefore recognizes the importance of European cooperation and decides to visit the French Prime Minister. The two politicians turn out to agree about numerous issues, which clears the way for the signing of the Treaties of Rome in 1957, which will prove to be of major significance for the developing EU [4]. In this case, too, a crisis turns out to produce an opportunity.

#### 5.4.1.3 Playing with Tight and Loose Couplings

Processes are usually characterized by a large number of relevant issues. Parties tend to couple and trade these issues. There is always the question whether these issues are coupled 'tightly' or 'loosely'. A tight coupling means that one issue cannot go without the other. The risk of such a tight coupling will be evident:

if one element fails, all other agreements will be cancelled as well. An advantage of a tight coupling may be that a party does not have the courage to torpedo all agreements by stubbornly clinging to its own interests. This would jeopardize its relations with others. A tight coupling may thus be a means to confront parties, often towards the end of the process.

The advantage and disadvantage of a loose coupling are exactly opposite. Issues are coupled during the process, but the coupling is loose and may easily be broken without too much damage as soon as one single issue obstructs progress with regard to the other issues. The disadvantage is that there are fewer opportunities to exert pressure on parties.

The process manager may play with these tight and loose couplings, and thus help the process forward.

In the history of the development of the EU, the 1970s are not exactly known as the most dynamic period. The EU finds itself in a stalemate—and not for the first time. Englishman Roy Jenkins is Chairman of the European Commission between 1977 and 1981. He intends to break the stalemate by adding a new dimension to the cooperation. He wants to entice the members to work more closely together with regard to monetary issues. This cooperation is highly contentious, and the debate becomes heated to an extent that other EU portfolios run the risk of being affected by the problems regarding this portfolio. Eventually, member states manage to establish the Economic and Monetary Union (EMU) by addressing the portfolio at the European level while decoupling it from the EU sensu stricto. The EMU becomes a 'hybrid, not entirely community, nor entirely outside it' [4]. Participation is open to EU members only, but it is not obligatory. The advantage of this arrangement is that disagreement about monetary cooperation does not obstruct other portfolios. Another advantage is that countries that oppose the EMU can no longer block process in this portfolio. Only the motivated countries remain, which of course makes it easier to achieve progress. There is still the possibility for other countries to enter the EMU if it proves to be successful—and also for this portfolio to be reintegrated into the regular European institutions at a later stage.

#### 5.4.1.4 Entering into Moderate Conflicts with Parties that have No Exit Options

Conflicts between the process manager and the parties are usually not conducive to the process manager's authority, but they cannot always be avoided. Much depends on which party is involved in the conflict. Parties with few exit options are likely candidates. They cannot exit the process following a conflict, and they have an interest in the process manager keeping the parties with many exit options in the process. This is why they will have, and need to have, a higher tolerance limit for conflicts with the process manager than the parties that do have exit options.

#### 5.4.1.5 Bypasses

A process manager may develop bypasses towards particular parties: relations in addition to those between the process manager and the party in the process. Such a bypass allows communication with a party along a second line. Bypasses make it possible to enter into a conflict with a party. After all, there are several relations between the process manager and the party (the 'bypasses'), permitting a more cooperative relation with the party in question apart from the conflict.

There is often such a bypass between the process manager and the representatives of the commissioning party. There is a line between process manager and process party, and a line between process manager and commissioning party. A conflict with the representative of the commissioning party may be mitigated through the line between process manager and commissioning party. The extra relationship may offer an opportunity to make the commissioning party somewhat sensitive to the dilemmas faced by the process manager, or to clarify the process manager's attitude position.

#### 5.4.1.6 Controlling External Support

Finally, if conflicts occur in a process, the process manager may seek the support of parties outside of the process. They may bring the importance of the process and of the process manager to the attention of the parties involved, which may provide these parties with an incentive to cooperate. In negotiations between business and civil society, a government may well express its support for the process, for instance, with the process manager pointing out that government measures will be taken unilaterally if the process fails.

### 5.4.2 *The Progress of the Process has an Independent Value…*

These actions allow a process manager to keep the discussion and negotiation process ongoing. From the perspective of classical management styles—command and control, substance and project management—this may hardly have any significance. The concepts addressed in Sect. 5.3—transforming intended substantive decisions into procedural decisions—will be little appreciated from this perspective. These management styles are aimed at taking effective, substantive decisions. Again, such an attitude stands little chance of success in a network: the stakeholders may obstruct such decisions. The rule tends to be: the firmer the decision, the stronger the stakeholders' resistance.

In case of such dependency on stakeholders, it is crucial for the manager to sustain the discussion and negotiation process—'keep the dialogue going' is the adage [11, p. 46]. Why is it so important to keep a process ongoing? What happens when a process is sustained?

#### 5.4.2.1 … Because Processes Result in Unfreezing

In the first place, parties may relinquish their established viewpoints; this is what we call unfreezing. Usually, such unfreezing can only occur as a result of repeated interaction with other parties. Such latent learning effects tend to be underappreciated because they fail to produce sufficient visible results. At the same time, they are an important breeding ground for later decision making. Negotiations are possible only when parties are prepared to reconsider and openly discuss their own views.

#### 5.4.2.2 … Because There Will be Relations and Opportunities for Gain

Secondly, relations develop between the parties negotiating in a process. These relations may be very appealing to the parties because they may serve to solve other problems than those on the agenda. Problems that have little or nothing to do with the process agenda tend to get solved 'in the slipstream' of the process. In addition, opportunities for gain may also emerge in relation to the items on the agenda. If this is the case, these parties will be enticed to arrive at decision making, because this will allow them to collect their gains.

#### 5.4.2.3 … Because the Cost of an Exit Option Will Go Up

Thirdly, the cost of an exit option keeps rising during the course of the process. After all, for many parties prospects of gain emerge during the process. A party that adopts a non-cooperative attitude during the process runs certain risks: it jeopardizes the gain for all parties and disturbs its relations with these parties. As a result, there may be substantial costs associated with non-cooperative behaviour. During the course of the process, the number of incentives for cooperative behaviour may increase, making it harder for parties to ignore the result of a process. The mild pressure exercised by the process itself thus results in more control than the harsh strategy of command and control.

The pressure to arrive at final decision making may also increase during the course of the process. Processes tend to have their own dynamics: once they have started and are in motion, it is difficult to terminate them prematurely.

Characteristic in this regard are the descriptions of the extensive negotiations between the ANC and the South African government about abolishing apartheid—the *negotiated revolution*. Although for a long time these negotiations are held in secret and hardly visible for anyone, the process develops its own dynamics. The white government used to believe that it could always exit the process, but finds at a certain point that this is no longer possible, although it has to sacrifice a lot more in the negotiations than it actually wants to. There are, for instance, young whites who are already anticipating the post-apartheid era, and thus refuse to leave the table. Others envision new commercial opportunities in case the international boycott is lifted. To part of the white delegation, the process is a matter of unfreezing: they drop their standpoints and reconcile themselves to the new reality-to-be. This puts so much pressure on the conservative parties in the process that they can neither halt nor leave the process [12].

#### 5.4.2.4 … Because There is Less Potential for Starting a Conflict

The advantages mentioned above will not always become apparent. The conflict of interests between the parties may be so severe that no prospects of gain emerge even during the process. Even in this case, however, keeping a process going may have an important fourth advantage. The fact that parties interact with each other reduces the risk of a public conflict. The parties may have an interest here: they know that the conflicting interests cannot be bridged, but they also know that the cost of a possible public conflict may be high. Keeping an interaction process going may then take on an independent meaning. This mechanism occurs for instance in peace negotiations: keeping a process going, even though there are no results, reduces the risk of a military conflict flaring up.

#### 5.4.2.5 … Because Outsiders' Expectations are an Incentive for Cooperative Behaviour

Fifthly, the progress of the process causes outsiders to have increasingly high expectations of the outcome of the process. This may be an incentive for the parties in the process to behave cooperatively.

> This is illustrated by the following example: civil society and business are negotiating about which type of package is the most environmentally friendly. Both parties may at some stage have an interest in delaying this dialogue: business because it fears that expensive packaging systems are the most environmentally friendly, and civil society because the outcome may be that they have spent years of campaigning for a packaging system that is less environmentally friendly than they thought.
>
> However, the process that they have entered into may raise high expectations among the general public. An often-heard complaint is that consumers are confused about the environmental profile of various packages. This may stimulate particular organizations to put pressure on the process: at least some results will have to be presented by the participants. The longer the process, the higher the expectations, and the more difficult it will be to use an exit option.

## References

1. Buchanan DU, Huczyncky A (1997) Organizational behavior: an introductory text. Prentice Hall, New York
2. Crystal J (2003) Bargaining in the negotiations over liberalizing trade in services: power, reciprocity and learning. Rev Int Polit Econ 10(3):552–578
3. De Bruijn JA, ten Heuvelhof EF, in 't Veld RJ (1998) Procesmanagement: Besluitvorming over de milieu- en economische aspecten van verpakkingen voor consumentenprodukten, Delft
4. Dinan D (1994) Ever closer union: an introduction to European integration. MacMillan, London
5. Fischhendler I (2004) Legal and institutional adaptation to climate uncertainty: a study of international rivers. Water Policy 6(4):281–302
6. Fisher R, Ury W (1981) Getting to yes. Houghton Mifflin, Boston

7. Hajer MA (1995) The politics of environmental discourse. Oxford University Press, Oxford
8. Hulse R, Sebenius JK (2003) Sequencing, acoustic separation, and 3-D negotiation of complex barriers: Charlene Barshefsky and IP rights in China. Int Negot 8(2):311–338
9. Innes JE, Booher D (1999) Metropolitan development as a complex system: a new approach to sustainability. Econ Dev Q 13(2):141–156
10. Kanter RM (1989) When giants learn to dance: mastering the challenge of strategy, management and careers in the 1990's. Simon and Schuster, New York
11. Kolb DM, Williams J (2008) Breakthrough bargaining. Harv Bus Rev Point 79(2):39–47
12. Sparks A (1995) Tomorrow is another country: the inside story of South Africa's negotiated revolution. Struik, Sandton
13. Susskind L (1999) Using assisted negotiation to settle land use disputes: a guidebook for public officials. The Lincoln Institute of Land Policy, Cambridge
14. Watkins M (2003) Strategic simplification: toward a theory of modular design in negotiation. Int Negot 8(1):149–167
15. Watkins M, Rosegrant S (2001) Breakthrough international negotiation: how great negotiators transformed the world's toughest post-cold war conflicts. Jossey-Bass, San Francisco

# Chapter 6
# A Safe Process: Protecting Core Values

## 6.1 Introduction

Chapter 5 addressed the open nature of decision making. Open decision making has major advantages, but it may also be quite threatening to the parties involved. They have particular interests and are not always sure whether their participation in an open decision-making process will actually serve their interests. They might get 'trapped', or perhaps open decision making will produce a result that they are not satisfied with.

The second core element of process management is therefore that it creates a safe environment for the parties: they should be certain that their core values will not be affected. In addition, the process should not be like a funnel trap, allowing only one direction and lacking a way back. A core value is a value that is of crucial importance to a party's existence. Harming this value means harming the party's essence; this inhibits its proper functioning. Section 6.2 will give some relevant examples.

We will elaborate the idea of a safe environment by answering four questions:

How can the parties' core values be protected? (Sect. 6.2)

What is the nature of the parties' commitment to the result of the process? Should they commit themselves to the result of the process in advance, at the start of the process? Is the result binding on the parties once the process has been completed and the parties have arrived at a result? (Sect. 6.3)

Should the parties commit themselves to the subresults that emerge during the course of the process? (Sect. 6.4)

How to deal with the exit rules of the process? How can parties be prevented from leaving the process prematurely, which would harm the progress of the process? (Sect. 6.5)

In this context it is relevant to note that the answers to these questions are largely counterintuitive. From a project perspective it seems likely to demand that the parties commit to the result of the process, that they do so as early on in the process as possible, and that they also commit to subresults and subdecisions

H. de Bruijn et al., *Process Management*, DOI: 10.1007/978-3-642-13941-3_6, 

during the process. Furthermore, parties should be given as little room as possible to use the exit option. If not, there is hardly any guarantee that a process will have a favourable ending, producing actual results that will be supported by the parties involved. In that case, a process may be a very laborious and inefficient management style.

It will be clear that such a project approach will be little effective in the process. The parties will perceive such a process design as highly threatening. They will not easily be prepared to participate in a process, and they will regard it as a funnel trap: it forces them in one particular direction and there is no way to escape. It is far more elegant to allow the parties room and guarantee that their core values will be protected under all circumstances. Room means that parties participate in a process in a relaxed way, allowing the process a fair chance of being effective. In the end the work should be done by the process itself, rather than by a set of predefined preconditions.

## 6.2 Protecting Core Values

Processes that fail to sufficiently protect the core values of the parties involved tend to have little chance of success. We will list some relevant examples here.

- The core value of a politically responsible politician (for instance a minister or a city executive) is his political accountability. At all times, he needs to be able to account for his actions vis-à-vis an elected body (such as the Parliament, or the municipal council). A politician who participates in a process may become wedged between the participants in the process on the one side and the elected body on the other side. After all, the result of the process may be a decision that fails to earn a majority vote in the elected body. It may therefore be little appealing for a politician to participate in a process; it is much more appealing for him if there are no strings attached. This will be different only if the process is designed so as to protect the politician's core value. An option for such a design is set out below.
- In many cases, companies are invited to participate in processes—such as in the examples addressing the environmental impact of packages and large-scale infrastructure. An important value for companies is the confidentiality of corporate information. They tend to be very hesitant to submit their figures about the development of particular markets or about the cost structure of their products to third parties. This is a core value for companies, which has to be protected in processes. Companies must be able to rely on the fact that they will not be forced by other parties to submit for instance their cost analysis of various packages during discussions on the environmental impact of packages.
- Societal organizations also have core values. The central mission of many societal organizations is to inform and to activate public opinion. Suppose a societal organization is invited to participate in a process on the condition that

during that process it will refrain from publicizing any information or views about the issue in question. Clearly, such a condition is unacceptable to many societal organizations. After all, one of their core values is to inform the public, and a process must never harm such a core value.

The fact that parties can be sure that the process will not harm particular core values may be a significant incentive for cooperative behaviour.

How can these core values be protected? The answer is simple: by establishing relevant process agreements—as will be illustrated by some examples later on.

The associated risk, however, is that the parties will too easily refer to their core values to protect their own position in the process. When companies refuse to submit any internal information, or when environmental organizations carry on a continuous campaign about the issues that are addressed in the process, chances for the process to succeed are of course slim. How to deal with this?

Processes nearly always result from negotiations between the parties (see Chaps. 3, 4). A party will put forward its core values during these negotiations. The other parties will learn that this party will only adopt a cooperative attitude if its core value is protected through process agreements. At the same time, the party with the core value may learn that other parties fear that this core value will be invoked too lightly. During the negotiation process, an arrangement will therefore need to evolve that protects the core values while preventing these from being invoked too lightly or too frequently. Let us take a look at an example.

Students in The Netherlands are allowed to travel by public transport for free, at the cost of the Ministry of Education. Every four years, the Ministry and the public transport companies (the Dutch Railways and the urban and regional carriers) negotiate about the cost of this student travel scheme. A key feature of these negotiations is the asymmetry in information: The Ministry of Education lacks information about the factual and expected use of the transport scheme (how many kilometres are travelled where and when?) and about the costs that the public transport companies allocate to the scheme. The Ministry of Education will need some insight into these details. However, the public transport companies may regard these details as confidential corporate information. There is, in other words, a core value that deserves protection.

If the parties then start to negotiate about a process for dealing with this core value, the outcome may be the following:

- The parties will initiate a joint study into particular facts and figures, such as the travel behaviour of students. The parties manage and supervise this study jointly in order to prevent the results from being influenced too much by the assumptions and views of one of the parties.
- With regard to facts and figures that cannot be obtained from research (for instance because they can only be derived from the corporate information of the companies), the parties make the following process agreement: facts and figures that one of the parties has available will be disclosed to the other party where possible. However, none of the parties is obliged to do so.
- If the public transport companies regard particular information as confidential, they will explain why this is the case. The Ministry of Education will form an opinion about this. If the companies regard particular information as confidential, the Ministry of Education reserves the right to collect the relevant information through other channels.

A first observation is that on paper, these agreements seem like a bureaucratic monstrosity. A process design, however, is the result of negotiations between the parties—so if the parties regard this agreement as useful, it is useful.

A second observation is that while the process agreement protects the core values of the companies, it also contains incentives to prevent these core values from being invoked too lightly. If the public transport companies use confidentiality as an argument, they will have to explain to their negotiation partner why this is the case. Of course the Ministry of Education may dispute their argumentation. By keeping open the option that the Ministry of Education collects the information itself, there is some incentive not to invoke this core value too lightly. Suppose, for instance, that the public transport companies refuse to submit information about some internal statistics. The Ministry of Education attempts to unveil these by itself, for instance by consulting with researchers who are knowledgeable about public transport tenders and thus about the cost structure of a transport service. This means that the public transport companies may well be faced with outcomes that they dislike, and that might prove to be a bomb under the process.

A third observation is that this process agreement was made prior to the process of substantive negotiation. Ideally, the process will be characterized by increasing trust between the parties and prospects of gain, eliminating the need for this agreement in the first place.

As will become clear later on, this is one of the characteristics of process management: there is a core value that is protected, but the process agreement is formulated in such a way that there are incentives to make only limited use of it during the process.

## 6.3 Commitment to the Process and to the Result

How is the commitment of parties to the process and its results affected by the protection of key interests?

### *6.3.1 Commitment to the Process Rather than to the Result*

A process ends with a result. This may be a package deal, for instance: a set of decisions that offer sufficient gain to each of the parties involved. Implementing these decisions will once more require cooperation between the parties involved; each party will have to meet its own obligations in order to guarantee the gain for the other party.

The first question is whether the parties can be committed to the final result at the *start* of a process. The answer to this question is simple. Processes have unpredictable dynamics, which makes it impossible to predict the final result of the process. This is the reason why the parties cannot be expected to commit to the result of the process beforehand. At the start of a process, the parties should always be offered room to distance themselves from the final result. If there is no such room, the process will be very laborious. This is because first and foremost, the parties will want to prevent the process from taking a direction that is unfavourable to them, and they will each want to influence any subresult in a direction that suits them best.

The parties' commitment will therefore necessarily be limited to a commitment to the process agreements: they are willing to join the process if it is conducted according to the rules that were agreed upon beforehand.

The second question relates to the nature of the commitment that parties make at the *end* of the process. At that point there is a result (for instance a set of decisions), which the parties must now implement. Again, a direct coupling between the result and its implementation poses a major threat to the process. If the parties know during the process that there is a direct link between the results and its implementation, there is a strong incentive for them to enter into close combat and be uncooperative. They will wish to influence the results to an extent that they will eliminate any risk that their interests might still be harmed in the implementation phase. This will result in very high decision-making costs, while the process might even remain without result. The parties need *room* in a process. A direct coupling between the results and their implementation does not offer such room.

### *6.3.2 Offering Room to the Parties*

The opposite idea is that of loose coupling [5]. This means that there is always some room between the results of the process and the ensuing implementation. Loose coupling may be an incentive for the parties to behave cooperatively in the process. It also reduces the decision-makings costs, because for the realization of their interests the parties are not fully dependent on the results of the process. They still have some room in the implementation phase, albeit to a limited degree. There is something counterintuitive about this as well. Intuition may tell us that an agreement is an agreement, which needs to be implemented without reserve. By agreeing that there is some room in the implementation, parties also create room in the process itself. We will use an example to illustrate the advantages of such loose coupling.

Suppose two organizations are planning to merge. They decide that this merger will be implemented as follows:

1. A study will be conducted into the opportunities and threats of a merger for both organizations.
2. Decisions about a merger will then be taken, based on this study: will there be a merger and, if so, how will this be accomplished?
3. These decisions will then be implemented.

The process described under (1) and (2) will be supervised by a process manager. It will involve the main stakeholders of the two organizations.

Suppose tight coupling is chosen: the decisions in phase (2) will be implemented exactly in phase (3); the research in phase (1) will strongly influence the decision making in phase (2). Although this might seem appealing, there may be two consequences.

Firstly, this strategy may place a heavy burden on phase 1. If—as a result of tight coupling—the results of phase 1 prescribe the implementation in phase 3 as almost compulsory, the parties will feel inclined to highlight all sorts of implementation problems in the research of phase 1. Phase 1 will likely be highly laborious, because the participants may come to feel that the party that loses the game in phase 1 has no further chances. Tight coupling may thus lead to stagnation.

Secondly, of course tight coupling does not take into account dynamics. During the process, there may be unforeseen and new developments, and the parties may learn. Consequently, to revert to the example above, the research of phase 1 will have lost part of its meaning in phase 3.

#### 6.3.2.1 Shape the Process in such a way that this Room is Used as Little as Possible

If parties are not required to commit to the results of the process beforehand, there is a risk that at the end of the process some parties will distance themselves from the results in which others already invested so much. The idea is, however, that a proper process constitutes a strong incentive for parties not to use this opportunity. During a process, a party will develop interesting relations and prospects of gain—and it may even learn to put its own views and core values in perspective. Distancing itself from the result at the end of a process would jeopardize all of these attainments. Although the process may be concluded without this party being tied to the result, the parties will, of course, meet again. If the parties have cooperated intensively and trustfully in a process, this distancing behaviour will be a heavy burden on these relations.

Once more, the dynamics occur that are characteristic for process management: offering sufficient room at the start of the process and investing in a proper and fair process may minimize the risk of the parties using this room. And once more, it is the process that should do the work, rather than a set of predefined preconditions. This is what often happens in processes: parties are offered room in advance, the process does its work, and at the end of the process it turns out to be difficult for parties to make use of the room offered, or—even better—they no longer feel the need to do so.

Decision making in the EU provides some interesting examples. A contentious issue in the EU, for instance, is the EU presidency. This issue was negotiated several times, including at a summit in Korfu in 1994. In an interview, one of the people involved, former Dutch Prime Minister Lubbers, analyses these negotiations. He confirms the rule that a new choice of president requires consensus: all member states should approve the candidate in question. However, as he points out: "Of course a small country is more likely to conform to the consensus." Again, there is a rule that inspires trust in the participants to the process. After all, they may as well commit to the process, because at the end consensus is required and therefore it seems as if the participants can never be forced to follow a direction that they are dissatisfied with: a member state may block the consensus and thus prevent an undesired decision being made. However, when it comes to the decision-making stage, it turns out that this room cannot always be used. Small countries in particular have limited possibilities in this regard.[1] Or, put more positively: large countries

[1] See the Dutch national newspaper *NRC Handelsblad*, 30-06-1994.

need not be afraid that any small country may block any decision just like that. This allows these larger countries to be a bit more inviting towards smaller countries than one would assume at first instance.

It might be added that the parties may establish additional process agreements about the conditions on which they are allowed to use the room offered to them. A familiar construction is that the parties are allowed to use this room if they can produce valid arguments for doing so. In other words, a party may distance itself from the results of the process, but only if it can make an acceptable case for this vis-à-vis the other parties. This reinforces the incentive not to use the room that is offered all too lightly: a party that rejects the results has to justify this, which makes it very clear to the other parties why the mutual relations that were developed apparently fail to stimulate this one party to commit to the results. We will elaborate this notion in a brief discussion of the core value of political managers.

### 6.3.3 Example: Process Management and the Core Value of Political Accountability

A manager with political accountability who participates in a process—either directly or indirectly through an official representative—is torn between two responsibilities:

- On the one hand he needs to be loyal to the parties in the process; he has gone through a process and achieved a certain result together with these parties.
- On the other hand he has an obligation to give account to an elected body (Parliament, for instance, or the municipal council). The members of such a body have an obligation to judge any proposals made by this manager with neither instructions nor consultations that support them in this task. Moreover, they are a democratically elected body and thus superior to the partners in the process.

This is the tension between what is called horizontal democracy—in the role of party, the manager engages in discussion and consultation with other societal parties—and vertical democracy: the manager has to give account to a superior, elected body. It is exactly this tension that explains why managers often avoid processes. The obligation to give political account may be regarded as one of the core values of the manager. Process agreements between parties will have to take into account this core value. A simple example of such a process agreement is the following:

- the manager participates in the process—either directly or indirectly;
- when parties achieve a certain result, they will offer the manager sufficient room to distance himself from this result;
- if the manager uses this opportunity, he will have to make an acceptable case for this vis-à-vis the other parties.

This simple rule is based on the familiar principle 'comply or explain': comply with whatever is agreed upon, or produce valid arguments that justify your resistance. The idea is that such a construction meets the manager's obligation to give account to an elected body. When a manager feels that he cannot get this body's approval of the result, he needs to have room to distance himself from a negotiated result. This way political primacy is respected. The higher the quality of the process—desirable results, proper representation of societal interests, strong players—the smaller the incentive for a manager to use this room. And: the harder it will be for him to justify the fact that he distances himself from the process. After all, when the process partners feel that the process was high-quality and that the manager is shirking the results too lightly, this may result in high costs for the manager. Future relations with partners in the process will be put at risk. The legitimacy of the process and the trust in the manager may be affected. Perhaps it is even fair to say that the better the process, the less likely it is that the elected body feels as if it has the opportunity to use its right to ignore the results of the process. Still, however, a process result that is good—at least in the eyes of the process partners—may be rejected by an elected body. After all, there is always a certain tension between horizontal and vertical democracy.

As a sidenote, in the example above it is the partners in the process who are the 'victims' of this tension—but the elected body may be the victim as well. The manager may for instance submit a certain measure to the elected body for approval, noting that the measure results from a lengthy negotiation process and, more importantly, that it is part of a package deal. If the elected body rejects the measure, this will be frustrating to the negotiation parties, who will then also refuse to meet their obligations in the package deal. In such a situation, the elected body hardly has any freedom of choice.

During the process, a manager may informally keep the members of the chosen body informed about the course of the process. He may do so substantively (which package deal is being negotiated?) as well as in a process-oriented way (which are the relations between the parties?). If the elected body takes these aspects into account, it may be in a better position to make a deliberate decision about the results of the process. How do these results relate to the process that the manager initially reported on? How much harm will be done to the partners in the process, and thereby to the position of the manager, if the elected body rejects the results?

## *6.3.4 The Position of the Initiator*

A party that initiates a process usually has a concrete reason for doing so. For instance, the party in question aims to realize an infrastructural project and knows that it depends on other parties in this regard. This party then faces a major dilemma.

- On the one hand, the party has to realize its project and thus invites other parties to participate in a process.
- On the other hand, these parties will only participate in this process if the initiator declares that the project is negotiable. Failing to do so would create the impression that the process only serves as a clever instrument to realize the project.

Let us take another look at the example of the Second Maasvlakte in Rotterdam. Rotterdam's ambition is to construct a 2,200-ha Second Maasvlakte, but gradually realizes that there is considerable resistance against this idea. Rotterdam will thus have to adopt some form of process management. It may invite other parties to join a process of consultation and negotiation. The Second Maasvlakte could be the subject of part of these consultations.

It is particularly the opponents of the Maasvlakte, who have obstructive power, who will probably demand that the idea of a Second Maasvlakte should to be open to discussion and thus be negotiable. If not, the process will be hardly appealing to them. They will get the impression that the process is only a clever instrument to get them committed to a Second Maasvlakte. Parties with managerial experience are aware of the fact that the power of a process is significant: relations and possibilities for gain will evolve, which make it very difficult for parties to reject the result of such a process—including a Second Maasvlakte.

Realizing a Second Maasvlakte is Rotterdam's main rationale for a process; Rotterdam would probably not start a process if it did not have this ambition.

Which arrangement could help solve this dilemma? Again, the most suitable arrangement is based on the idea of loose coupling between the result of the process and its implementation. This arrangement will look as follows:

- The initiator declares that the envisioned project is negotiable.
- The other parties acknowledge that the project is a prerequisite for the initiator to join the process. They offer the initiator the possibility to distance himself from the results at the end of the process, if and in so far as the initiator makes a valid case for this vis-à-vis the other parties.

Again, this arrangement offers room to the initiator: he can distance himself from the result of the process if it fails to include the envisioned project. The initiator has an exit option. In return, at the start of the process he declares the project negotiable. Two scenarios are possible now:

- At the end of the process, the initiator finds that the result is not satisfactory to him. He may then distance himself from the results. In considering this, he will make a trade-off: on the one hand, he would be jeopardizing the positive results of the process (the package), but on the other hand he would then have the freedom to try to realize his project in a different way.
- At the end of the process, the initiator finds that the results are in fact satisfactory to him. This may mean that he has realized the envisioned project, at least to a sufficient degree. Alternatively, he may not have realized his project, but he is nevertheless pleased with the results of the process. After all, various opportunities for gain have evolved during the process, and some of these possibilities were unforeseeable at the start of the process. Moreover, he may have learned during the process that there are significant alternatives to the envisioned project, for instance, or that there is so much resistance against the project that it has no reasonable chance of success.

If Rotterdam declares that the Second Maasvlakte is negotiable and then engages in a process with a number of stakeholders, Rotterdam might be highly pleased with the results at the end of this process, even if the Second Maasvlakte that is agreed upon is much smaller than the envisioned 2,200 ha. Rotterdam may have learned that the resistance to a larger Maasvlakte is so strong that it cannot reasonably be realized. There may also have been negotiation during the process, offering Rotterdam new, unexpected possibilities for gain. Suppose one of the stakeholders (a small harbour town called Vlissingen) proposes that the Rotterdam Port Authority should cooperate closely with the Vlissingen Port Authority from now on. And suppose that there is a new technological development: there is a new type of ship that has a considerably greater container capacity than its predecessors, but which cannot reach the Belgian port of Antwerp, Vlissingen's competitor, because of its substantial draught. As long as the international treaty obligation to deepen the Westerschelde River—which provides access to Antwerp, but runs through Dutch territory—has not been fulfilled, these ships would have to divert to Vlissingen. This would call for a rapid and strong development of Vlissingen. Although the actual decision making has taken a different course, one can imagine that this package would be highly attractive to Rotterdam. A strong position for the Rotterdam Port Authority in Vlissingen could then be exchanged for a smaller Maasvlakte, while Vlissingen could create a distinct profile for itself as a strong competitor of Antwerp thanks to the new technological developments.

#### 6.3.4.1 The Process Should Do the Work

The message may become repetitive, yet it cannot be emphasized enough: the process should do the work. Predefined preconditions hamper the process. Trust between parties and prospects of gain can only evolve through a process. Mutual trust and prospects of gain may have important implications for parties' core values:

- perhaps parties will develop more respect for the core values of others. They understand that a company cannot simply publicize all of its figures, that an environmental organization should be allowed to stir up societal unrest, and that a manager needs the support of an elected body.
- perhaps the parties will put their own core values in perspective: if there is distrust, companies might be more willing to submit certain figures, societal organizations might be more willing to show some reserve towards the press, and a manager might be more willing to seek a confrontation with an elected body.

The process should do its work: beforehand, core values are respected and no commitment to the result is required. The commitment will evolve during the course of the process, while the process may promote respect for the core values of others, and stimulate parties to put their own core values in perspective.

### *6.3.5 Incompatibility and Opportunistic Use of Core Values*

To conclude this issue, we will briefly address two questions regarding the notion that parties' core values should be respected:

- How to deal with mutually incompatible core values? What to do if the protection of the core values of party A harms those of party B?
- What to do if both parties show strategic behaviour? For instance if they define their viewpoints as core values and thereby declare them non-negotiable?

The following example illustrates a situation in which both problems emerge.

Suppose two countries engage in peace talks 15 years after ending a war. During this war, country A has conquered a plateau, which it continues to occupy. Country A argues that holding onto this plateau is one of its core values: national security would be threatened if the plateau were handed over to country B. Country B, on the other hand, argues that the retrocession of the plateau is a precondition for its participation in the process. The integrity of its own territory is one of its core values that cannot be compromised.

Now suppose that an eminent politician or diplomat acts as a facilitator or process manager in these negotiations. This facilitator will be faced with two problems:

- The core values of the one party harm the core values of the other party.
- The process manager feels that these are not actually parties' core values, but views that should be negotiable.

As far as the latter problem is concerned: the example shows that the process manager is unlikely to succeed if he tries to convince the parties that these are not core values. Core values tend to be core values when a party regards them as core values. Respecting them usually creates more goodwill than attempting to make a party abandon them.

The arrangement that may be designed for situations of this kind will always have to seek to give room to the parties. Processes should offer security: parties must be convinced that the process does not threaten what they define as their core values. If such room is offered, parties may be willing to join the process.

This room may be offered through an arrangement such as the one set out above: offer the parties the possibility to distance themselves from the results of a process. The process should then be allowed to take its course: creating relations and producing possibilities for gain. Once it is no longer appealing to parties to use the possibility of distancing themselves from the results, the process manager may attempt to move onto the decision-making stage.

Of course we have no solution for the kind of stalemate described in the example above, but it is clear that the essence of every solution comprises two parts:

- It is necessary for these parties to join the negotiating process. Only in a process of mutual interaction can they learn how to put their own views into perspective and develop relations with each other and with other parties (superpowers, neighbouring states, financial institutions), and only then can they learn that they have possibilities for gain: more economic growth, security, support from financial institutions, good relations with (regional) superpowers. This does not mean that a process always leads to unfreezing—it would be too naïve to expect that—but it does mean that no unfreezing will occur if there is no process at all. Moreover, as long as the process continues, the chance of an armed conflict is smaller than if there were no process.
- The parties will only join a process if their core values are protected. The promise that this will be the case offers them the room to join the process. This, too, may be realized by offering the parties the possibility to distance themselves from the results if their own core values are not sufficiently protected. The process agreement may stipulate, for example, that the parties have the right to distance themselves from the final result and

leave the process if it does insufficient justice to their views. This language may sound vague, but it means that country A is allowed to leave the process if it cannot hold onto the plateau, just like country B is allowed to leave the process if it does not retrieve the plateau. These are conflicting promises, but this is something that a process manager is supposed to be able to deal with. The process should do its work: it should produce good relations and gains.

The example illustrates one way to deal with conflicting core values: if they are substantively incompatible, they can be defined in a process-type way. In the above example, the agreement will not be that party A will always be allowed to retain the plateau, but rather that party A may distance itself from the results of the process if it feels that these do insufficient justice to its views about the position of the plateau. The same agreement can be made with party B.

In general, such arrangements can be made with parties that define their own views as core values.

The dynamics of this negotiation process may look as follows. There is an agenda with numerous issues—cooperation on water, transboundary cooperation with regard to security, tourism, demilitarization, family reunion and so on. All of these issues represent prospects of gain. If a number of other parties (superpowers, regional powers) also have an interest in a peace treaty, they may expand this list of prospects of gain: economic support, knowledge exchange, trade agreements. Parties need to develop trust in the process and in each other, and they need to see prospects of gain; therefore it is wise to create ‘quick wins’ (as will be explained in the following chapter) and to address issues that are relatively simple and that quickly produce prospects of gain. The complicated issue of the plateau should perhaps *not* be addressed for the time being. Sometimes, managing equals postponing.

At a certain moment there is a draft package deal that, ideally, offers both parties sufficient gain. A prospect of gain and an intensive process may promote mutual trust. What may happen next? Perhaps trust and prospects of gain will allow parties to put into perspective what they regarded as a core value at the start of the process—retention or retrieval of the plateau, respectively. If this is not yet the case, some pressure may induce them to put their views in perspective: if they do not move at all, the entire package deal will fail. Moreover, the other countries will be highly disappointed, which will affect the relations with these other countries. In many cases such processes end ‘in a pressure cooker’—such as in Camp David, where a process manager’s pressure compelled either a ‘yes’ or a ‘no’ with regard to a final decision.

One thing needs to be added to the example above. Especially in the case of peace negotiations there tends to be a deep-rooted distrust between parties. This means that such processes are highly likely to fail. *Ex post* reconstructs of successful processes, however, consistently reveal the elements outlined above.

## 6.4 Postponing Commitments During the Process…

A large number of decisions are taken during a process. The question is how strongly the parties have to commit themselves to these interim results. The intuitive answer tends to be that such commitments are necessary. This funnels the decision making (certain options are excluded, for instance), it brings to the surface at least some initial results, and it slowly but steadily forces parties into a certain direction. As we said earlier, this might seem like a decisive strategy, but in reality it may be threatening to the parties in a process. If a party is required to commit to subdecisions during the process, there is a significant chance that there will be a stalemate between the parties because they perceive the process as a funnel trap in which every subdecision represents a point of no return.

An alternative to this approach is the principle that commitments may be postponed: when there is a substantive subdecision, parties are either not required commit to it entirely, or they are offered the possibility to change their minds if they do. Application of this design principle prevents the perception of a process being like a funnel trap. It offers room to the parties involved, and is thus suitable for a process characterized by uncertainties and distrust.

For that reason we can draw the same conclusion as in Sect. 6.3: offering room is necessary for the progress and the quality of a process. A process manager who primarily strives to make decisions and solve problems as quickly as possible will be unable to cash in on a number of positive effects of this design principle.

### *6.4.1 ... Reduces the Decision-Making Costs*

Processes tend to have a multi-issue agenda, and multi-issue decision making processes are always partly unpredictable. During the process it is unclear which final decisions will be made. If a substantive decision is submitted in such a situation, it will probably be impossible for the parties to foresee the implications of these interim results for the final decision making and for their own position.

This may lead to a kind of decision making in which the parties want to secure their positions as much as possible. As a result, lengthy negotiations may be needed to ensure that the decision does maximum justice to the parties' interests. In turn, this may cause the decision to be either complex and detailed, or sketchy and vague. Postponing commitments may avert such laborious decision making.

An additional step is to offer parties a veto-right option: they may postpone commitments, and if they dislike the final result they may veto it.

A familiar principle, particularly at the start of processes, is that parties agree that decisions can only be made unanimously. In other words, all parties have a veto right. At first glance such processes may seem powerless and unlikely to produce results. But on second thoughts, it will often be this agreement which gives parties the courage to engage in a process. They know that they can always say no, which means that their core values cannot be harmed. Without this agreement, the process might not even have been possible at all.

At the same time parties will realize that they may compromise their own position if they consistently use their veto right or threaten to do so. The alternative is to negotiate with the other parties about the conditions for not using the veto right.

The veto right is a feature of decision making in the EU. Centred on this feature, several institutions have developed that are conducive to the progress of the decision making. It may seem as if every country has a veto right, but its use is bound to specific, though unwritten rules. To name an example: if at a certain moment a large number of countries support a proposal, it is inappropriate for a single country to obstruct the decision making [1, pp. 58, 59]. Marjolin, a former vice-president of the European Commission, has made such a rule explicit. The European Commission operates by the 'golden rule of not taking any action... likely to encounter an outright veto that would have left no room for negotiation' [3, p. 56].

Such subtle codes were used during the GATT negotiations as well. Robert E. Hudec wonders whether the replacement of GATT with the WTO—which has a more formal connotation—has in fact benefited the decision making. He particularly questions the greater legal formality of the WTO.

'The secret of getting one's way under a regime of consensus decision making is the ability to make joining consensus a better answer than the alternatives. In GATT, there was always the risk that the larger countries would forge ahead on their own, stone-walling adverse dispute settlement decisions if it came to that. In the WTO, it is likely that large governments will feel more constrained to follow dispute settlement decisions, partly because such decisions will automatically be "legally binding", and partly because the larger governments have made a greater political commitment to their own constituents to provide an effective dispute settlement system. Thus the greater "legality" of the WTO may invite the smaller countries to stick to their guns longer and more forcefully'.[2]

These and similar forms of postponing commitment keep the decision-making costs low. If each of the issues required a binding decision, negotiation processes would probably be very lengthy and laborious due to the uncertainty about the implications for the final decision making.

## 6.4.2 ... *Offers Possibilities for Dealing with the Decision-Making Dynamics*

It may be added here that in complex decision-making processes a large number of decisions are taken that will turn out not to influence the final result. What may seem like a principal issue at the start of a process may become a mere detail after some time. The opposite is also possible: in retrospect, a detail may prove to have been a crucial decision.

In the context of an unpredictable decision-making process, offering room is a contingent design principle. On the one hand, it prevents much energy being spent on settling differences of opinion that will later prove to be mere details. If, on the other hand, irreversible decisions are taken in the opening phase about alleged details (which are thus likely to provoke less resistance), these may condition the rest of the decision-making process in an undesirable way.

---

[2] Comments by Robert E. Hudec on Jackson [4, p. 224].

### 6.4.3 … Offers Possibilities for Building Mutual Trust

Process designs are made to bring the parties together. In such a context, distrust is almost unavoidable.

If a decision has to be made against a background of distrust, parties tend to start an investigation into the interests and hidden motives of the other parties. Why does a party hold a particular view? What is a party's hidden agenda? Rumours and suspicions may prevail, which can be a serious threat to the process. If the parties are uncertain and/or suspicious, though not forced to take a position because they can postpone commitments, trust may evolve—both in the process manager and in the other parties.

In short, postponing commitments is a vital stimulus for mutual trust and thus for a successful process. It may gradually reduce uncertainties and strategic behaviour. A process manager who offers room invests in a cooperative attitude of the parties. Paradoxically, offering room to strategic behaviour may cause it to diminish after some time (Chap. 8 will address the problematic nature of the concept of strategic behaviour).

### 6.4.4 … Stimulates Learning Processes

It is important that a process should largely be a learning process shared by the parties (!). During the process, new insights will become available, facts will turn out to be different from what is generally assumed, and even normative views can change. Making binding commitments at an early stage may seriously hamper such learning processes.

As we pointed out earlier, parties tend to have no common frame of reference at the start of a process. When these parties are brought together in a process, the first thing they need to do is put in perspective their conviction that they are the only ones who are right. Only then can they build up a common frame of reference. This calls for a learning process, which can be realized in particular by the room created by postponing commitments. A frame of reference comprises a number of dominant ideas among the parties about the course of the process, its substance, its interim and final products, and so on.

As a result, postponing commitments usually leads to learning processes and the building up of a common frame of reference behind a façade of 'non-decision-making'. Such a frame of reference may be conducive to rapid decision making later in the process.

If a country has to make an important infrastructural decision—for example about its future air traffic infrastructure—there will usually be an organized dialogue between the interested parties. This dialogue will result in a particular outcome: the parties may have reached agreement about particular issues, while they have improved their ability to formulate their differences of opinion about the remaining issues.

In such dialogues, one of the main points of discussion tends to be the tension between economic and ecological interests. The process manager may strive for the parties not only to show 'participative openness' (i.e., to exchange their viewpoints), but also 'reflective openness': they are willing to bring their own views and values up for discussion [6]. This might break the stalemate between economic and ecological interests on a number of points.

'Reflective openness' implies, however, that the parties involved may have to abandon some of their deep-rooted views, which is not without risk. After all, their opponents might use these overtures for their own purposes ('environmental movement in favour of airport expansion'; 'industry admits economic advantage is unclear').

Suppose the parties have agreed beforehand that the result of the dialogue will have a direct impact on the political decision making. In other words, there is a tight coupling between the dialogue and the decision making. This may make the parties hesitant to bring up their own views for discussion. After all, there is a chance that these will be abused during the political decision making. As a result, the dialogue may not be characterized by 'reflective openness', but merely by 'participative openness'. If there is an opportunity to postpone any commitments to views and decisions, on the other hand, 'reflective openness' may evolve.

With regard to learning effects, it is important that a process should proceed from substantive variety towards selection (as has been explained in Sect. 3.6 in Chap. 3). During the starting phase of a process, a variety of options should be considered, allowing an authoritative selection to be made during the final phase. A development from variety to selection cannot be reconciled with overly quick decision making at the start of the process.

### 6.4.5 … *Takes the Pressure Off the Decision Making*

During the course of the process, the results have to be formulated, made presentable and communicated. By definition, open decision-making processes generate a large quantity of information, the volume of which has to be reduced in the course of the process.

This implies that the aggregation level at which the available information must be processed will rise sharply in the course of the process. This will decrease the relevance of various subdecisions. Moreover, external communication of the results also requires a certain simplicity and unambiguousness, which in the end of the process is an incentive for the parties to reach decisions.[3] This is another reason not to make substantive decisions too early in the process: given the aggregation level at which the information will eventually have to be processed, substantive decisions may have little impact on the final result.

Table 6.1 summarizes the information above. It also lists the consequences of the opposite attitude, which is aimed at making a maximum number of 'binding' decisions in a minimum amount of time.

---

[3] See description of Burger et al. [2].

**Table 6.1** Comparison: room versus preconditions

| Design principle: process manager offers room | Design principle: process manager formulates preconditions and aims for consolidation |
|---|---|
| Helps to create trust between the parties | Helps to create distrust between the parties |
| May diminish strategic behaviour | May intensify strategic behaviour |
| Improves the progress of the process, which is a value in itself | Increases the risk that the process may be aborted, which may greatly hamper the start of a new process |
| Processes are started through the generation of variety; subsequently, room will improve the quality | Processes are started through the generation of variety; decision making can be conditioned in an undesirable way. Decision making implies unnecessary loss of managerial energy |
| The process manager may get to know the parties' sensitivities and thus estimate the scope for win–win packages | The process manager loses his touch with the parties and is increasingly ineffective in fulfilling his role |
| Decisions can be made in a relaxed atmosphere and lessons can be learned | Decisions are taken in an uptight atmosphere, which is hardly conducive to decision making |

Of course, a major risk of this design principle is that the decision making does not produce any visible results for quite some time. Chapter 7 will discuss strategies to deal with this risk.

## 6.5 The Exit Rules of the Process

An important design principle is that a process should have exit rules. The process agreements may stipulate, for instance, that after some time the parties may consider whether they wish to continue their participation in the process. This may be an important agreement that lowers the threshold for particular parties to join the process.

At the same time, the process manager will do his utmost to prevent parties from actually leaving the process. Ideally, the process gradually becomes so satisfactory to parties that leaving is no longer an appealing option for them. Here, too, the parties must be offered room, but the process must be so appealing that they will not be inclined to use this room. Some additional observations may be made here.

### *6.5.1 The Participation Paradox: An Exit Option may be Appealing*

The aim of involving parties is to improve the quality of and the support for the envisioned decisions. Paradoxically, the opposite may be accomplished. Particular parties participate in the process and thus obtain more and better information than if they had not participated. They may use this information at the end of the process to oppose the decision making rather than to support it. The information obtained can enable a party to put up better and perhaps more convincing resistance to the decision making than if the party in question had not participated.

A party that displays this behaviour at the end of a process may be accused of opportunism. This is why it may be attractive for a party to use the exit option during the process and then oppose the decision making. This will decrease a party's risk of being accused of opportunism (after all, it uses one of its rights), while it is still able to use the obtained information to oppose the decision making.

## *6.5.2 Threatening to Leave: Double Binds for the Process Manager*

When a party threatens to leave the process, the process manager is faced with several risky double binds:

- *vis-à-vis the party that wants to leave*. The party that wants to leave the process will wish to do so because the process has too little to offer. If such a party wins and is allowed to leave, the process will be harmed. If such a party loses, this reinforces its conviction that the process has little to offer, which may also harm the process.
- *vis-à-vis the other parties*. The other parties will closely monitor the actions taken by the process manager. If the process manager loses and the party leaves the process, the credibility of the process manager will be harmed. If the process manager wins, the first double bind may manifest itself here too: the losing party believes even more strongly that the process has little to offer. This will be different only if the process manager can offer this party some prospect of gain, which, however, may cause the other parties to believe that a threat to leave apparently pays off. This is not conducive to the progress of the process either.

### 6.5.2.1 An imminent exit can hardly be managed: it is up to others to do this

It is important to realize that the process manager is hardly in a position to manage an imminent exit; after all, the option to leave is a right that the participants have. This is where another mechanism of process management can be used once more: frame problems as a conflict between parties, rather than as a conflict between the process manager and one of the parties.

There is a major risk that the latter will happen anyway, because a party's leaving harms the process manager's key interest: it jeopardizes the progress of the process. In such a conflict, the process manager will always be faced with the above-mentioned double binds, which makes any choice made by the process manager highly risky.

However, if the conflict can be framed as a conflict between the parties, the process manager can once more play an independent role as a facilitator in the conflict. In so far as the process manager can influence the conflict, he will do so indirectly: through other parties in the process or through the process environment.

## References


1. Andersson M, Mol APJ (2002) The Netherlands in the UNFCCC process. Leadership between ambition and reality. Int Environ Agreem 2(1):49–68
2. Burger J et al (2001) Science, Policy Stakeholders, and Fish Consumption Authorities: developing a fish fact sheet for the Savannah River. Environ Manage 4:501–514
3. Dinan D (1994) Ever Closer Union: an introduction to European Integration. MacMillan, London
4. Jackson JH (2000) The role and effectiveness of the WTO dispute settlement mechanism. In: Collins SM, Rodrik D (eds) Brookings Trade Forum, pp 179–212, 220–231. The Brookings Institution, Washington, DC
5. Perrow C (1984) Normal accidents: living with high-risk technologies. Princeton University Press, Princeton
6. Senge P (1992) The fifth discipline: the art and practice of the learning organization. Doubleday/Currency, New York

# Chapter 7
# A Process with Sufficient Speed: Incentives for Progress

## 7.1 Introduction

So far, we have argued that process management is characterized by openness. The main stakeholders are invited to participate in a process and are involved in drawing up the agenda. Openness, however, is not without risk for these stakeholders. They can perceive the process as a funnel trap: once they have joined, they may feel that they are forced in a certain direction without being able to leave the process. It is therefore important that parties' core values are protected. For the sake of these core values, parties are offered room at crucial moments. For instance, they are not required to commit to the result of the process beforehand, and they are offered an exit option.

The design principles 'openness' and 'protection of core values' inevitably prompt the question: what guarantees sufficient speed in the process? After all, in the worst case there will be many parties presenting many agenda items, which makes the process difficult to manage. If these parties also put forward many core values, which they declare not to be negotiable, there is every risk that the process will be very slow or even end in a stalemate.

It will be clear that networks hardly offer any room for classical, project-type mechanisms aimed at increasing speed. Establishing a deadline, for instance, provides an incentive for speed in a *project*, but in a *process* this may stimulate resistance. After all, some parties have no interest in decision making before the deadline expires, as this harms their interests. They will therefore delay the decision making. Command and control hardly stands any chance either, since all parties are mutually dependent.

The main incentives for progress in a process evolve when a process is appealing to the parties. A process must be sufficiently appealing at the start (partly as a result of the work of the process architect), but new possibilities for gain should also evolve during the process. Based upon what has been described in the previous chapters, these dynamics may be summarized as follows:

H. de Bruijn et al., *Process Management*, DOI: 10.1007/978-3-642-13941-3_7,

- *An appealing start.* Parties are invited to participate in a process and are involved in drawing up the agenda. They will place items on the agenda that are appealing to them. Other parties will do the same thing, which results in a 'multi-issue' agenda. This multi-issue agenda makes a process appealing, because it permits couplings between the various items. The process is also appealing because the core values of the parties are protected.
- *Forming relations.* Relations will then evolve between the parties in the process. These relations can promote decision making about the items on the agenda. The relations can also be used, however, to discuss entirely different subjects that lie outside of the scope of the process. Parties tend to solve many problems in the slipstream of the process.
- *Prospects of gain.* As the process progresses, possibilities for gain will present themselves to the parties. They identify opportunities to realize the ambitions they placed on the agenda. They may be unable to realize these in full. However, gains may be made even then, because they have learned that there is insufficient support for particular issues among the other parties.
- *Unfreezing.* An interaction process usually also contributes substantially to parties' unfreezing. Parties put their own views, which may have been rigid at the start of the process, in perspective. This is highly significant, because this enables the parties to exchange issues for—or to couple them to—other issues. As long as the parties' views remain rigid, a process will only result in losses. If parties put their own views in perspective, however, they will be more easily convinced that certain decisions or package deals will benefit them.
- *Gains are paid out late.* To protect their core values, parties are allowed to postpone their commitment to the interim results of the process. As a result, a large number of problems that are difficult to solve may still be on the table at the end of the process. At the same time, postponing commitments means that parties still have opportunities for gain towards the end of the process. After all, postponing commitments to interim results means that the gains that particular parties expect from these interim results have not been paid yet. This is in compliance with one of the golden rules regarding processes: parties' gains should never be paid out too early, because this may eliminate important incentives for cooperative behaviour.
- *Postponing commitments and offering exit options.* Paradoxically, offering exit options may speed up a process. Parties are offered a possibility to reconsider their commitment at a later stage, under particular conditions. This room may remove their hesitation and stimulate them to agree with a particular proposal.

With regard to the earlier-mentioned process 'More space for rivers', the final list of forty projects will be included in a plan with a heavy, formal status. However, it turns out to be difficult to convince all relevant politicians to approve the list of forty projects. The State Secretary then announces that although these forty do constitute the list, it will be possible—even after the adoption of the plan—to exchange projects for other projects, under the condition that these new projects are hydraulically equally efficient as the cancelled projects, and that they do not entail any financial or planning-related setbacks. Under these conditions, the politicians are willing to adopt the list.

- *End: profit and loss account.* As a result, there will be a stage at which the parties have developed relations and prospects of gain, while there are still a number of problems that are difficult to solve and that fail to evoke consensus. Each party will then draw up a profit and loss account. On the positive side of the balance are the relations developed and the gains collected, on the negative side there are the losses and the unsolved problems. For particular parties, who have no interest in the problem, the latter side is uninteresting; for others, who have an interest in a particular solution of this problem, it represents a form of loss.
- *Profit and loss balance positive for a critical mass: speed.* The speed of the process will increase if the profit and loss account shows a positive balance for a critical mass of parties. They wish to collect their gains and therefore to make final decisions. At this point there will be an important psychological mechanism: parties tend to anticipate on collecting their gains, which increases their urge to speed up the process.

It is clear from the above, however, that the end of a process is difficult to predict. Ideally, the parties should have some degrees of freedom in establishing the deadline of a process. They should be able to control the timing of the process in mutual consultation, allowing for a conclusive decision to be taken when there is momentum for it; in other words, when there are a sufficient number of positive profit and loss accounts among a sufficient number of parties to arrive at final decision making.

This would be in an ideal situation, but of course reality tends to be different. This is why we will discuss some additional concepts here.

What to do if there are insufficient possibilities for gain, at least in the parties' perception? The process manager may manage the process in such a way that new incentives for cooperative behaviour continue to be created (Sect. 7.2). The process manager may make use of the staffing of the process (Sect. 7.3). The speed may be increased by playing with so-called 'quick wins' (Sect. 7.4). The way a process is organized may increase the speed of the process (Sect. 7.5). In conclusion, Sect. 7.6 will discuss the possibilities of providing steering in a process through command and control. Although a command and control style conflicts with the idea of process management, situations may occur during (and as a result of) the process in which command and control may succeed.

## 7.2 Incentives for Cooperative Behaviour

A process manager has four different options to create incentives for cooperative behaviour during a process. He may do so through:

- the architecture of the agenda;
- the planning of activities;

- interventions by third parties, which may reframe conflicts or provide these with multiple dimensions;
- offering parties repeated opportunities for realizing their interests.

The idea is that these incentives will promote the process dynamics described above. We will further elaborate these incentives using the following example.

Parties A and B are negotiating about the most suitable location for a country's national airport. There are three different options:

- the current location, which is close to a metropolis;
- Flyland, an airport at sea, which is also relatively close to the metropolis; and
- a peripheral location, far away from the metropolis, in an area that has much space available.

A study is conducted into the effects of building the airport at each of these three locations, focusing, among other issues, on the costs, the economic and ecological effects, and the technical opportunities and risks. The parties are requested to submit feasible and well-argued proposals for a choice of location.

Party A prefers to reach conclusions and arrive at decisions quickly, and takes the view that the economic effects are paramount. Party B, on the other hand, values a careful study and careful decision making, and feels that the ecological effects are most important. It will be obvious that these are the ingredients for a stalemate. How can incentives for cooperative behaviour be created in this process?

### *7.2.1 Architecture of the Agenda: A Balance Between Productive and Obstructive Power*

As has been mentioned before, the agenda of a process should be a multi-issue agenda whenever possible. Only then can parties exchange and couple issues and will strong incentives for cooperative behaviour evolve. It is also crucial for the agenda to be drawn up in such a way that it provides an incentive for all parties to use their productive power. Productive power is positive. Parties use their resources in order to create. The opposite of this is obstructive power: parties use their resources to obstruct decision making.

For party B in the example above, there is hardly any incentive to use its productive power. This may be different if the agenda is framed in another way. For example, the agenda may address the question how the area that becomes available if the airport is relocated may be used to strengthen the ecological structure of the metropolitan area. In other words, process managers should always consider ways to activate the production power of the main stakeholders.

### *7.2.2 Planning of Activities*

A second incentive for cooperative behaviour is created by drawing up an intelligent timeline for the relevant activities.

#### 7.2.2.1 Incremental Planning

At the start of the process, ambitions are purposely set low. The agreed targets are modest, which makes it easier for parties to accept the agreement. This agreement could be supplemented with process rules that describe how developments will be monitored. If the problem turns out to be more serious than previously assumed, it may be necessary to take an additional step. In some situations it will appear to be possible to reach consensus about mechanisms aimed at increasing the targets if the monitoring shows that the problem is more serious than previously assumed [13, p. 137].

#### 7.2.2.2 Sequential Planning of Activities

If party A feels that decisions should be made quickly and party B prefers making them carefully, the following structure may be planned: round 1 will serve the interest of party A and round 2 that of party B.

This may create an incentive for party A to give maximum consideration to the interests of party B already in round 1. After all, the more these interests are taken into account, the greater the chance that the decision making in round 2 will be quick.

> The process manager may propose, for instance, that a quick scan will be made of the economic and ecological effects of the choice of each of the locations. Once the results of this quick scan are available, the parties can form an initial judgment about these effects and formulate additional questions. These questions are to be answered in a new study.
>
> This proposal acknowledges the interest of quick decision making: a quick scan is conducted, rather than a detailed study. It also meets party B's views, however. If the quick scan should leave party B to feel that this scan lacks quality, it may demand an additional study. This may then be an incentive for moderate behaviour on the part of party A. The fact that a quick scan is made does not undermine the need for sufficient quality. Otherwise, there will be much room for party B to demand an additional study.

The agreement thus starts out by addressing the interest of party A. However, A is aware of the fact that B's interest will be served in the next phase. This is an incentive for party A to take into account party B's interest. If party A fails to do so, party B will have a strong incentive to use its obstructive power, which will cause a delay.

#### 7.2.2.3 Parallel Planning of Activities

In the above example, there is an incentive for moderate behaviour by party A in particular. This is different when, at the same time, sequential connections are made for various activities. The structure may be as follows: the activities serving the interests of party A are conducted; both party A and party B are involved in this. At the same time, the activities serving the interests of party B are conducted; here, too, both party A and party B are involved.

Suppose party A is particularly interested in the economic study into the new Flyland, while party B is more interested in the ecological study into the alternative utilization of the current metropolitan airport. The economic and ecological studies may be connected in parallel. The fact that the studies in which both parties are interested are conducted in parallel provides incentives for cooperative behaviour. A party that is unreasonably critical of a study that serves the interest of the other party will invite the same attitude in the other party towards the other study. This mechanism may stimulate parties to adopt a moderate attitude.

This example serves to illustrate the effect of parallel connections between different activities that meet the interests of different parties: there will be an incentive for cooperative behaviour.

It should be noted that when it comes to substantive agenda setting and planning of activities, the process manager is guided by the incentives that the agenda and the planning provide—rather than by substantive considerations.

As a sidenote, the process manager should always continue to think actively. He should not apply these strategies mechanistically. After all, parties may always show strategic behaviour. If a sequential planning is applied, party B may for instance be overly demanding, only to make extra demands in round 2.

#### 7.2.2.4 Intervention by a Third Party: Assigning Multiple Dimensions, or Reframing

A significant incentive for cooperative behaviour may be provided by an intervention by a third party. The idea is that such a third party may assign extra dimensions to a conflict or reframe the conflict, thus creating additional room for negotiation.[1]

For this purpose, a process manager may invite parties to formulate a mutual conflict as accurately as possible and then submit it to a third party. This third party is to pass judgment on the conflict. Agreements may be made regarding the implications of the third party's judgment. These may range from 'parties will accept the judgment of the third party in advance' to 'parties will take the judgment into consideration in their further consultations'.

The latter agreement may seem to be rather powerless, it but may nevertheless prove meaningful in practice. After all, the judgment does not necessarily imply that any of the parties is right. It may also imply that the third party is pointing out that multiple dimensions are relevant in the conflict, which may create new room for solutions between the parties.

[1] See for instance Field [5], in which the Hartford case is particularly relevant. Research Parker and Wragg [10] points in that direction. The establishment of a counternetwork that is positioned opposite the existing network strongly affects the decision making. See also Huygen [6, p. 136], Rein and Schon [11, p. 8] and Lakoff [9, p. 9].

Parties are at conflict about whether the construction of an island in the sea at a particular location will be harmful to the marine ecosystem, since it may alter the current. Party A believes that the present plans can do no damage, but party B feels that the risk of damage is too great. The third party may pass a judgment on the variables that affect the current, such as the location, the place where sand will be extracted to build the island, and compensatory technical measures. This third-party intervention may allow the process between A and B to continue. The discussion is no longer a simple yes/no question (will there be ecological damage or not?), but it now has multiple dimensions. For instance: where and how will sand be extracted with a minimum of ecological damage? Which compensatory technical measures are needed? The fact that the stalemate now has multiple dimensions creates additional room for negotiations.

The third party may also reframe the conflict: it captures the conflict in a different vocabulary, which creates room for negotiations. The first form of reframing is reformulating a negotiation conflict into a research question. Again, a process manager will invite parties to formulate their conflict as accurately as possible. With the help of a third party, this conflict is then translated into a research question, which is submitted to the independent third party

Parties are at a conflict about the safety of Flyland. Party A has become convinced that the large number of seabirds may seriously threaten the safety of incoming and departing aircraft. Party B disagrees.

They may reframe this conflict as a research problem. This first of all requires acuteness on the part of the parties; they have to formulate a number of research questions, such as:

- Where and when do large seabird populations congregate?
- How regular and predictable are the movements of these birds?
- How will the construction of an airport affect the behaviour of these birds?
- Which measures can be taken to chase birds away?
- How do birds react to which measures?
- Which are the uncertainties in this study?

These questions may then be submitted to researchers, in accordance with a protocol drawn up by the parties. This protocol may stipulate, for example, that in answering the above questions, the researchers also indicate which research findings are 'solid' and which of them contain uncertainties. The result may then help the parties forward in their negotiations. The research may show, for instance, that the birds' movements are difficult to predict, but that there are many technical possibilities to keep birds at a distance. These solutions, however, are costly, which creates new room in the discussion: it is not just about whether the plan is 'safe or unsafe', but also about the amount of money the parties are willing to invest in risk-reducing measures and about the cost-effectiveness of those measures.

In brief, the idea is that the research conducted by a third party may facilitate the negotiations. Of course, the fact that parties *can* use this arrangement is at least as important as its actual use. If a party knows that a particular viewpoint may be the subject of study at some stage, this may be an incentive for this party to adopt a moderate attitude and to do maximum justice to the facts when taking a position.

This type of intervention transforms a conflict into a research question. A variant to this theme is that parties transform a research conflict into a negotiation question.

There may for instance be uncertainty about the yearly migration of the birds. Do they always follow the same route, or may there be significant differences between the annual routes? This lack of data will call for several years of research. The resulting data, however, will never be completely reliable—and yet these data are a prerequisite for party B to grant its approval to Flyland.

An ongoing conflict about the research may result in a stalemate. Reframing the issue as a negotiation question may imply that parties A and B assume that a worst case scenario applies to the birds' migration, and that they will therefore seek technical possibilities to keep the birds at an acceptable distance from the airport. They will start to negotiate about this, which bypasses the research question.

## 7.2.3 Repeated Opportunities to Realize One's Own Interests

Processes are usually characterized by a multi-issue agenda. Eventually, the parties will have to reach a decision about each issue. It is important that the process design should offer all parties multiple consecutive opportunities to realize their interest with respect to this issue. This gives parties confidence regarding the integrity of the process design, and it also prevents parties from stirring up a conflict regarding every single decision. After all, if parties have only one opportunity, they will go all out to make full use of it. This will be a driver for conflicts in the process.

A process agreement may for instance imply that:

- Parties commission research institute A to study the impact of birds on the safety of the airport-at-sea.
- If the parties feel that this research lacks authority, they may subject it to a peer review by research institute B.
- If, given the review by peer B and the subsequent reaction to this by research institute A, the study still fails to fully satisfy the parties, they can commission additional research.
- Once there is a decision about the impact of birds that is supported both by research institute A and by the parties, the research is concluded. If new circumstances present themselves later, this may be a reason for further research.

These agreements may then be supplemented with certain conditions, for instance that the support of a minority of the parties will suffice for arrangement (2), whereas a majority is necessary for arrangement (3) and a qualified majority is required for arrangement (4).

To outsiders, this process agreement may seem cumbersome. However, it is understandable from the point of view of a party that has major interests in the process and that is afraid that research institute A might produce a biased research report. The essence of the agreement is that it offers this party repeated opportunities if the report does indeed turn out to be biased. Moreover, if a number of repeated opportunities are available *for each issue*, a party has sufficient room to realize its own interests. Here, too, it is important to distinguish between these repeated opportunities being offered and being actually used. The process will have to take its course (also see Chap. 6): if relations between parties and prospects of gain develop during the process, incentives will evolve to make only limited use of the room offered.

In conclusion, it is recommended that a process offers repeated opportunities. This stimulates moderate behaviour, as parties realize that they will meet again in a following round, and that reasonable behaviour in the current round will benefit their position in the next one.

This mechanism also applies to processes in which there is uncertainty about the seriousness of certain facts, and in which scientific research may offer some certainty at a later stage. 'Parties can exploit this uncertainty in order to stall progress.' [4]. The speed of processes may be increased if they are designed in such a way that new information may be incorporated at a later stage, allowing for evaluations to be nuanced and for decisions to be revised. Eckley refers to this as 'dependable dynamism' (Ibid.).

## 7.3 The Process is Heavily Staffed

The idea behind the need for heavy staffing has been described in Chap. 3: it promotes the external authority of the process and the commitment of the parties involved.

### *7.3.1 Heavy Staffing Creates Opportunities for Gain and Incentives for Cooperative Behaviour*

It should be added that a heavy staffing also provides extra opportunities to generate gains and incentives for cooperative behaviour. After all, 'heavy representatives' have extensive networks at their disposal, which increases the appeal of participation in the process: parts of the extensive networks of the other parties become available to a participating party.

Heavy representation also provides more opportunities to conclude win–win package deals. After all, heavy representatives have more room to negotiate, as they are less troubled by consultation with those they represent. This is one of the reasons why it is easier for them to accept a loss than for a light representation, which is tied by consultation to those represented.

From 1990 to 1994, there is a process of decision making regarding the Per+ project, a large investment project carried out at Shell Netherlands Refinery. This decision making is strongly interactive: Shell involves the main stakeholders in the decision making. Two key figures play a role: a member of the Royal Shell Group's committee of managing directors, and a politician representing the province in which the project is to take place. 'They have, each in their own area, organised and consolidated the required commitment within their organisation.' [15, p. 5].

Another advantage of extensive networks is that multiple issues can be included in a package. This is because they offer extra possibilities of coupling problems and solutions. However, these advantages are counterbalanced by a number of potential risks associated with a heavy representation. A heavy representation implies that the participants are difficult to manage, precisely because they have these networks of relations. In other networks, participants may have other interests, which are at conflict with a proper course of the process. These participants may then use their powers against the process: they may for instance use their networks to influence the process via third parties.

Consequently, this *paradox of heavy representation* results in the fact that an important determinant for a successful process (high-level participation of the main stakeholders) is also the main threat to the process (the heavier the representation, the greater their destructive power).

This paradox requires some form of management by the process manager. This is not easy: heavy representatives tend to be difficult to manage. The essence of such management is that the representatives will be dependent on each other also in other situations and in the future.

> In the above example of the Per+ project, Shell is interested in a good investment climate in the area, and thus in good relations with the government. The government, in turn, will increasingly and more frequently have to rely on the self-regulation of companies when it comes to its environmental policy, and therefore it has an interest in good relations as well.

This may have a moderating influence. The role of the process manager is the same as the one described in the previous paragraph: he needs to identify and specify these dependencies, hoping that they will automatically have an impact during the process.

## 7.4 Quick Wins

### *7.4.1 The Threat of a Low Product/Time Ratio*

An important problem for the process manager is that outside pressure can only be generated if the outside world is regularly faced with results of the process. If the parties in a process fail to generate products, outsiders may get the impression that the process is nothing but sluggishness. This may compromise their faith in the process. Interactive processes are slow—perhaps too slow, in the eyes of the outside world.

The criticism of the low product/time ratio may be refuted by pointing out that the products that are eventually delivered demonstrate that the decision making as a whole was not sluggish. Time is a relative concept: time tends to be lost at the start of the process, and this loss is made up for at the end of the process.

All the same, there is a significant problem here. It may only become clear *after* the process that the product/time ratio was high. *During* the process, however, the process manager will have to live with a low product/time ratio while being unable

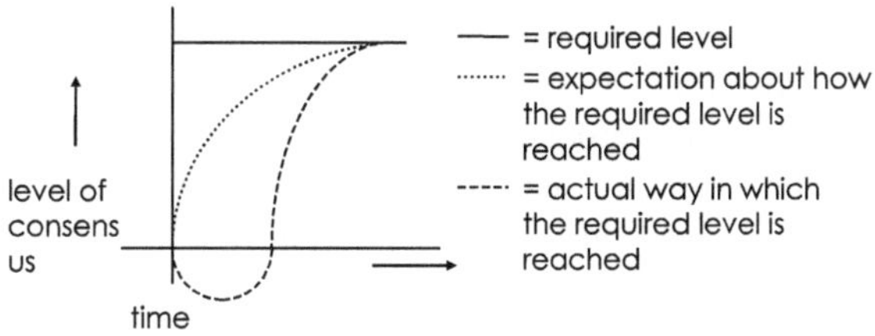


**Fig. 7.1** The incubation period of a process approach

to fully guarantee that he will complete the process successfully, which would allow the product/time ratio to rise.

These problems may occur especially at the start of a process. A process is based on interaction: parties exchange views, negotiate and learn. As a result, unfreezing may occur: during the design process, parties become less certain about their views, information and/or aims. This is a positive development. Room can be created for negotiation and decision making only when parties learn how to put their own views into perspective.

However, unfreezing may also be a problem for a process manager. Those represented may start to believe that the process approach will be unsuccessful: unfreezing takes time, there seems to be no progress, and a representative questioning his own views may also be a problem for those represented. In short, it looks as if the process is insufficiently effective. Figure 7.1 shows a schematic representation of this risk.

The dotted curve shows the course of a process of change as desired by the parties: from a situation of limited consensus to one of sufficient consensus to arrive at decision making.

When the curve falls below the $x$-axis, parties may get the impression that a process approach does not work: there is too much divergence and too little progress. Actually, the process has some sort of incubation period, in which unfreezing takes place. This incubation period is a prerequisite for a proper process, but it also makes it vulnerable. It is the process manager's role to make clear to the parties in the opening phase—which is therefore a critical phase—that the process does observe the intentions of a process design.

Another important point in this regard is the fact that parties learn during the process. Black-and-white pictures adopt a shade of grey, original problem definitions lose their meaning and are replaced by more fruitful problem definitions. Such a learning process may be one of the objectives of the process, but it is also a threat. Particularly among outsiders, who experience these learning processes just as intensively, the impression may arise that there is no progress whatsoever. This may threaten the legitimacy of the process. The implication is of course that the outside world should be informed about the learning processes that parties go through during the process. In fact, outsiders should be involved in this process. However, this also increases the need to show quick wins: results that are contributable to the process and that increase its legitimacy.

In other words, a process is vulnerable, particularly in the beginning. The process manager may therefore have to ensure that there are interim products that show that the process is advancing and that it is more than the procession of Echternach, in which participants took three steps forward followed by two steps backward. Quick results may promote the participants' support for the process as well as its legitimacy.

Which are the options for achieving quick results? One is to place the items that will evoke the least resistance high on the agenda. This strategy is recommended particularly for processes that are known to be lengthy [8]. The answer to the question *when* an issue should be dealt with and *when* a process should result in a product thus becomes mostly process-based: when it is desirable for the sake of support for the process.

Initiatives are being taken in many countries to set up 'smart cities'. These are districts with an advanced information infrastructure, offering a multitude of services. Setting up such a district requires the cooperation of many parties (municipalities, residents, providers of infrastructure and services, and so on), who are facing many risks (administrative, technical and economic ones). Setting up a smart city therefore calls for a process of consultation and negotiation between these parties. This process should result in the parties making a joint estimate of the risks, and also a joint plan that ensures that the smart city is organized in such a way that it is sufficiently attractive to all, or at least most, of the parties.

When it comes to the design and management of such a process, it is of major importance that preliminary results will present themselves regularly. This is an important incentive for parties' cooperative behaviour. It assures external financers that the process is productive, and thereby it is a parameter that increases the speed of the process. Without such interim results, chances are that a process will stagnate or that the momentum for a smart city will disappear [16].

Alternatively, the process manager may attempt to entice the parties to arrive at decision making more quickly and to act more quickly. The actors who are willing to speed up are rewarded. At least some actors will go along with this, which will increase the speed of the process as a whole and may create a new élan.

In the earlier-mentioned programme 'More space for rivers', a list has been made containing 700 projects. A definitive list of projects is to be selected on the basis of this list. The government has made 1.9 billion euros available for the implementation of these projects.

It is of course substantively and administratively challenging to arrive at a definitive list, although some projects will clearly be eligible in any case. In order to prevent stagnation, the responsible State Secretary introduces the concept 'trendsetter project'. A project that is awarded this status may start already, in anticipation of the adoption of the definitive list. This status has two significant advantages. Firstly, trendsetter projects will receive much political attention. Secondly, these projects are first in line for the distribution of the 1.9 billion euros, and advantage that is difficult to ignore. Evidently, a substantial number of projects apply for the trendsetter status, which increases the speed of the process. Third parties notice that there are actual developments, which creates goodwill for the process.

## 7.5 Conflicts are Transferred to the Periphery of the Process

This design principle has already been discussed in Chap. 3, which explained how and why conflicts are kept as far away from the centre of decision making as possible. After all, due to the conflicting interests of the parties, there may be limited potential for conflict resolution at the centre of the project. If there are too

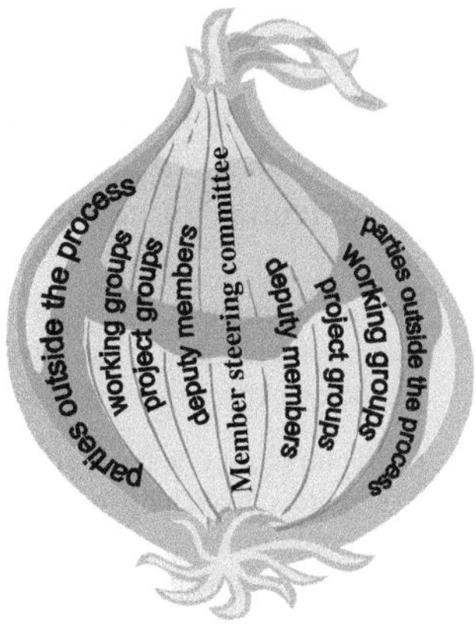


**Fig. 7.2** The various shells of a process

many conflicts at the centre, there is a risk that substance displaces the process: the process stagnates, preventing the advantages of keeping a process going (see Chap. 5) from materializing.

The organization of a process may be used to reduce conflicts. There will always be a number of individuals who form the core of the process and individuals who are positioned more towards the periphery of the process. Suppose the structure of the process comprises a steering committee, project groups and working groups. The core of the process then comprises the members of the steering committee, who represent the parties at a high level and who have to take decisions together. In the next shell, there may be deputy members, for instance. The project groups are one shell further out. They prepare the steering committee's decisions. The penultimate shell contains working groups, while the outer shell contains the parties that are outside of the process, but nevertheless take or have an interest in it (see Fig. 7.2).

The process manager may prevent conflicts in a process from being sucked into the centre by framing them in such a way that there is no need for them to be solved by the parties at the centre of the process. Sibenius aptly refers to this as a 'non-inflammatory conference structure' [13, p. 142]. The following strategies may be used in this regard.

- *Frame conflicts in such a way that they have to be settled outside of the process.* The most convenient situation is the one in which conflicts have to be solved by actors outside of the process. After all, this means that the negotiating parties do not have to take any particular position, and the conflict places no burden on their mutual relations.
- *Frame conflicts in such a way that they have to be settled in the outer shell of the process.* If the first strategy is impossible, another option is to submit a conflict to a body outside the centre of the process. The process manager will have to ensure that there is no need to solve all conflicts in the steering committee. In some cases, it is useful to formulate them in such a way that they have to be solved by the project groups mentioned above.

- *Frame conflicts in such a way that they can be solved at the level of deputies.* An important feature of heavy representatives is that they often work with deputies. A framing of conflicts that allows deputies to solve most of them eases the pressure on the relations between the parties. After all, this construction leaves less potential for conflict among the direct representatives of the parties.

Deputy members can play an important role in decision-making processes. Sparks mentions a number of the advantages in his description of the negotiating process between the South African government and the ANC [14, p. 31, pp. 78–79]:

- Deputies may help in creating some understanding of a party's room for negotiation;
- Deputies may help when relations between representatives are difficult, for instance because mutual prejudice prevails among the parties;
- Deputies constitute additional channels for disseminating and receiving information.

- *Frame conflicts in such a way that the coalitions of supporters and opponents are always different.* Of course, there will always be a number of conflicts—particularly in the last round of the process—that can only be solved in the inner shell. In the example, this is where the members of the steering committee are positioned. An important rule for the process manager is that it should be avoided that the same coalitions evolve in these conflicts time and again. Problems can be formulated in such a way that the coalitions are not too predictable and that they may change a number of times. If the coalitions were always the same, a block might form at the centre of the process, which may seriously jeopardize the process.

Does all of this signify weak leadership and insufficient decisiveness? Again, the answer should be that there is little chance of success in a network if a style of command and control is chosen. It is more intelligent to give the process a chance by reducing conflicts, which will allow it to play its positive role. Letting the process take its course (see again Chap. 5) creates a breeding ground for decision making and even for command and control (see the following section).

## 7.6 Command and Control: Both a Driver and a Result of the Process

In conclusion, processes aim to promote cooperation between the parties. Good cooperation is conducive to the speed of the process. Cooperation can be distinguished from two other drivers, which may also increase speed: (1) the exercise of power through command and control; and (2) competition. Therefore it is risky if cooperation is the sole focus of a process manager. After all, if a person knows that his opponent wishes to realize his goals through cooperation with him, it is very tempting for him to refuse to cooperate. As a result, the opponent's dependency on this person will increase, and the opponent may have to sacrifice more to win the person's support.

This is why it is important to make use of the other two drivers for speed, besides cooperation: power and competition.

This section will briefly describe the need and possibilities for combinations of these three drivers.

We therefore distinguish between three types of drivers[2]:

- Drivers resulting from power in a hierarchical structure
- Drivers evolving between competitors in a market-like structure
- Drivers resulting from cooperation in a network-like structure.

### 7.6.1 Power

In a number of situations there is a power relation between actors. In a hierarchical structure (for instance hierarchy in an organization, or norm-adressees vs. their law) there are dominant and subordinate actors. Besides these obvious kinds of power there are more subtle variants. For instance, one of the actors involved in a process has a large amount of resources available that allow him to force other actors into cooperating.

The counterpart of power is the sanction. This power is effective—it influences other parties' behaviour, because they aim to avoid potential sanctions.

Power, however, has its limits and its disadvantages. Actors with power may force other actors into cooperating, but apparently they still need this cooperation. This is where there is a limit to power. After all, actors will always have some amount of room to determine whether and how they will cooperate with the actor who exercises his power.

The use of power evokes counterpowers. Actors try to avoid the word of power. They hide, they cause delay, or they combine forces to organize resistance. Time and again, it seems as if power, particularly power per se, organizes its own counterpowers.

### 7.6.2 Competition

Competition unleashes many—and strong—powers. Companies that are aware that they are each other's competitors will aim to perform better than the other party. They will innovate and/or provide their products and services cheaper, among other things. Public institutions may also compete with each other. In case of overlapping authorities or responsibilities, public organizations may end up in a process of trying to outperform each other, which may lead to innovations and improved service. In public as well in private contexts, actors who sense the

---

[2] See also Williamson [17] for comparison.

presence of a competitor will tend to make a little extra effort. This driver, too, may be used to advance processes.

However, there is also a downside to competition. Competition, as Schumpeter [12] already highlighted, also implies destruction. Wherever there is competition, multiple parties will do the same thing, at least for some time. The efforts of the losing party will be lost. This simply qualifies as a waste.

### *7.6.3 Cooperation*

Actors who are aware of their mutual dependence will cooperate as long and in so far as they see opportunities to realize their interests. The structure that generates cooperation is that of a network.

Cooperation does justice to the dependencies between actors. If the cooperation process works out well, actors are given both room and opportunities to complement each other and to arrive at agreements that are beneficial to all of them.

Cooperation has its downside too, however. If parties are offered every room to form coalitions, reach agreements and develop synergies, this may easily result in an endless process of searching and talking, in which increasingly complex kinds of cooperation are constructed. This makes it very difficult for new actors to join the process, and there are few incentives for progress.

Table 7.1 summarizes the above.

Fruitful processes tend to be subject to a combination of the three drivers. The process manager will therefore aim to use all three drivers. A classical hybrid, for instance, is *coopetition* [1]. Parties that are in coopetition with each other combine processes of cooperation and competition. On the one hand, they cooperate and complement each other in order to be able to provide the complex products and services demanded by clients. On the other hand, they are in competition with each other. The coalitions that they form when they cooperate are temporary and fragile, and will never develop into a better functioning relationship. Parties are aware of that, and they will therefore be on their guard even when they cooperate. They will keep looking for improvements that are interesting for them. Cooperation and competition may also succeed each other in time, or run in parallel.

**Table 7.1** Three types of drivers that may be used in processes

| Power | Competition | Cooperation |
|---|---|---|
| Structure | | |
| Hierarchy | Market | Network |
| Positive outcome | | |
| Quick, unambiguous | Innovation, efficiency | Synergy |
| Risk | | |
| Generates strategic behaviour | Destruction | Sluggishness and stagnation |

Another familiar combination of drivers is that of cooperation and power. This is where the term *throffers* is used: threats (pressure, power, command and control) that go hand in hand with offers (overtures, cooperation).

All three drivers have played a role in the process in which companies have been quite successfully encouraged to develop and use packages with a better environmental profile. Successful package development requires much cooperation—primarily between companies that wish to package their goods and companies that develop packages. However, additional cooperation is needed, especially with regard to the reduction of negative environmental impacts. Packages are developed, produced and processed in long business chains. Steps taken in one link of the chain affect the possibilities to make environmental improvements in other links. The companies in the different sections of the chain therefore need to cooperate.

In addition to this cooperation, there is a lot of competition. Firstly, there is competition with regard to materials. Glass, plastics, paper/carton and metal, to name four important materials, are all in the running for becoming the packaging standard for the products that they can hold. In the case of milk, for instance, there is competition between glass, plastic and carton—but also between the companies that produce the same product. The various beer-producing companies are each other's competitors, but at the same time they cooperate with regard to packaging. They understand that there are multiple reasons for standardization of bottles and crates. Cooperation is therefore unavoidable. In short, competition and cooperation are relevant at the same time: there is a need for *coopetition.*

Although the new packages are developed during voluntary processes, through agreements and *coopetition*, these processes will never pick up any speed without the threat of a government that is present at the background and that is expected to intervene through legislation in case the business sector does not produce sufficient results voluntarily.

In summary, in this process all three drivers have been relevant, in some kind of equilibrium.

We will conclude with a number of examples of pressure versus cooperation, or, to use the language of this book, of command and control versus process. It should be noted that in these examples, it is either the process that drives command and control (strategies 4 and 5), or command and control that drive the process (strategies 1, 2 and 3). In other words, these two drivers are bundled [2, 7].

*Strategy 1* Command and control may be a *driver of a process of cooperation* because it puts pressure on the parties in a process. Suppose there is a process going on with effective dynamics: relations are being developed and possibilities for gain evolve. The speed of the process may be increased, however, through the use of various kinds of command and control.

A minister who is engaged in a process to negotiate with the business sector about an agreement, but who is threatening to introduce unilateral rules at the same time, is likely to reach a better agreement and reach it sooner than a minister who opts for a process only. The same phenomenon occurs in the relations between states in the form of 'bulldozer diplomacy': negotiations in a process are accompanied with a display of power in order to speed up the negotiating process.

A notable effect of these and similar kinds of command and control is that it gives parties a different perception of their gains. These gains include not only the results they can achieve by negotiating in the process, but also the prevention of the threat emanating from command and control.

In the example of the negotiating minister: preventing unilateral rules may be a form of gain. If the business sector is to cash in on these gains, it will have to conform to the minister's wishes in one way or another. There is a reasonable chance that this will speed up the consulting and negotiating process.

*Strategy 2* Command and control can be a process driver if it is used while room for a process is offered at the same time.

The Board of Management of an organization with the structure and the culture of a network may unilaterally announce a merger between two divisions and simultaneously offer room to these two divisions: in a consultation process, they are allowed to exert a strong influence on the strategy and structure of the new, merged division. The divisions are thus presented with a trade-off: resisting the merger decision (negative energy) or making optimum use of the room offered (positive energy). If they opt for resistance, this will require an almost impossible kind of management. After all, there will always be units within the division that opt for using the room offered and that will therefore be difficult to manage.

If the Board of Management confined itself to designing a process for achieving a merger, some of the consequences would be easy to predict: there would be reactive behaviour on the part of the divisions, attempts to delay the process, no loyal participation in the process, and so on. Here, too, the combination of command and control and process management may ensure a certain speed of the process.

*Strategy 3* Command and control may be a driver if it is used to install in the parties a sense of urgency regarding the need for a process. As we pointed out earlier (Chap. 4), a process only has a chance of success if the parties feel such a sense of urgency.

An often-heard complaint is that a particular initiative starts off as a project and then degenerates into a process. A design for a project is made, which helps to activate the parties. They find the design in conflict with their interests, and thus oppose its realization. Such a development is regarded as undesirable and may be a reason to recommend that the initiative in question should be developed as a process rather than as a project: invite the stakeholders to set up a project during the process. However, there is something inevitable about the development from project to process: there is no process without a project. The project is the driver for parties to become active; if there is no project, parties will usually not be interested in committing themselves to a process. Put differently, the project installs in the parties a sense of urgency: the project teaches them that a process is needed for them to arrive at a decision for which there is broad support.

The inevitable (and, occasionally, tragic) nature of this development may be used actively. The manager may propose a detailed project, not in order to realize it, but rather as a driver for a process. This is a form of command and control.

*Strategy 4* Command and control may be helpful when at some point a critical mass of parties stands to gain by the process. After all, these parties will wish to cash in on their gains, and have an interest in completing the process quickly. They will put pressure on the other parties—either intentionally or unintentionally, either explicitly or implicitly—to complete the process, and may use for this purpose the networks of connections that were built during the process.

This may suddenly accelerate the process. This is the reason why many processes end ‘in a pressure cooker’: there is a sudden acceleration of the process because a critical mass of parties wishes to arrive at decisions. This pressure cooker may give rise to some special dynamics, but these tend to be difficult to predict.

- Potential losers who try to obstruct the decision making in the pressure cooker may receive generous compensation from the parties that stand to gain. After all, these will want to reach decisions quickly and will therefore tend to give in more quickly.
- Potential losers who try to obstruct the decision making in the pressure cooker may be put under severe pressure by the other parties as well. This changes their perception of gain: the losers want to prevent their obstruction to the decision making from harming their relations with the potential winners; they will therefore be more likely to interpret a particular decision as a gain.
- It is also possible to take process-based rather than substantive decisions about the issues that are of great importance to the potential losers (‘further consultations will be held about issue x’; ‘the parties decide not to take any actions regarding issue y without the approval of party A’, and so on). In many cases, this means that the parties continue their interaction in another process, with a new agenda that is sufficiently attractive to the potential losers to help complete the current process. This is what is called a *roof tile construction*: the final round of a process is designed as the first round of a new process. Thus, various decision-making processes are coupled in an overlapping fashion, causing the end of each decision-making process to be influenced by the next decision-making process. Dixit and Nalebuff formulate this as follows: ‘To avoid the unraveling of trust, there should be no clear final step. As long as there remains a chance of continued business, it will never be worthwhile to cheat. So when a shady character tells you this will be his last deal before retiring, be especially cautious’ [3, p. 158].

*Strategy 5* Command and control is potentially fruitful when a process has failed. If a manager asks a number of parties to arrive at decisions in a process and they are unable to do so, there will be room for unilateral decision making. After all, the parties have learned that they are unable to solve this particular problem in mutual consultation. They have learned that a process entails high decision-making costs, which is one of the reasons why they more inclined to accept unilateral interventions. The process has thus created a breeding ground for command and control.

A process may fail without the parties’ intent or awareness. Alternatively, parties may enter the process feeling confident that it will fail no matter what. In the latter case, the following two situations may occur:

- *Situation 1.* A party aims to implement a strategic plan and is convinced that the stakeholders will be unable to reach a decision by mutual consultation. Based upon this conviction, the party decides that the strategic plan must be implemented. This provokes resistance among the stakeholders, but after some time they will give up their resistance and accept the implementation of the plan.
- *Situation 2.* A party aims to implement a strategic plan and is convinced that the stakeholders will be unable to reach a decision by mutual consultation. Nevertheless, this party gives them the time and opportunity for a process. This process fails, which makes the stakeholders more amenable to a unilateral decision and creates room for the manager to announce the implementation of the strategic plan. His intervention is so strong that the stakeholders accept his decision.

The party in question must decide which of the two situations will result in the lowest decision-making costs. At first sight, the second situation is a waste of time, but it may be efficient to allow the parties some time to learn that they are unable to reach a decision in a process. Parties who merely resist a decision (situation 1) will not undergo this learning process. The question is therefore which costs are higher: those of a failed process (situation 2) or those of the parties' resistance (situation 1).

## References

1. Brandenburger BJ, Nalebuff AM (1997) Coopetition. Bantam Doubleday, New York
2. De Bruijn JA (2005) Roles for unilateral action in networks. Int J Pub Sect Manag 18(4):318–329
3. Dixit A, Nalebuff BJ (1991) Thinking strategically. The competitive edge in business politics and every day lifes. Norton, New York
4. Eckley N (2002) Dependable dynamism: lessons for designing scientific assessment processes in consensus negotiations. Glob Environ Change 12(1):15–23
5. Field CG (1997) Building consensus for affordable housing. Hous Policy Debate 4:801–832
6. Huygen J (1995) Culturele en strategische dimensies in besluitvorming: Techno-politiek en het GBA-project. In: 't Hart P, Metselaar M, Verbeek B. Publieke besluitvorming. Den Haag, VUGA, pp 125–148
7. Koffijberg J (2005) Getijden van beleid: omslagpunten in de volkshuisvesting. Over de rol van hiërarchie en netwerken bij grote veranderingen. IOS Press, Amsterdam
8. Kotter JP (1995) Leading change: why transformation efforts fail. Harv Manag Rev 73(2):59–67
9. Lakoff RT (2000) The language war. University of California Press, Berkeley
10. Parker G, Wragg A (1999) Networks, agency and (de)stabilization: the issue of navigation on the River Wye, UK. J Environ Plan Manag 42(4):471–487
11. Rein M, Schon DA (1986) Frame-reflective policy discourse. Beleidsanalyse 15(4):4–18
12. Schumpeter JA (1934) The theory of economic development. Harvard University Press, Cambridge
13. Sebenius JK (1991) Designing negotiations toward a new regime. The case of global warming. Int Sec 15(4):110–148
14. Sparks A (1995) Tomorrow is another country: the inside story of South Africa's negotiated revolution. Struik, Sandton

15. Van den Bosch FAJ, Postma S (1995) 'Strategic stakeholder management: a description of the decision-making process of a mega-investment project at Europe's biggest oil refinery. Shell Nederland Raffinaderij BV. Rotterdam. In: Management Reports Series 242, Erasmus Universiteit/Rotterdam School of Management, Rotterdam
16. Weening HM (2001) Het vliegwiel vervlogen? Een evaluatie van het verloop en de aanpak van het proces rond de Kenniswijk, Delft, commissioned by Dutch Ministry of Transport, Public Works and Water Management
17. Williamson OE (1975) Markets and hierarchies: analysis and antitrust implications. The Free Press, New York

# Chapter 8
# The Process Manager and the Substance of Decision Making

## 8.1 Introduction

The fourth core element of the process approach is substance: the process that is developed under the guidance of the process manager must be sufficiently substantive. After all, a process without substance is empty.

The preceding chapters have already pointed out repeatedly that a decision-making process may degenerate into a process for the sake of the process. This may affect its speed (core element 3), but also its substance. When a process drifts too far away from the substance, it is vulnerable and fails to meet its original objective: a process is designed to produce substantive problem definitions and problem solutions.

This chapter describes how the quality of the substance in a process can be protcctcd. Scction 8.2 examines the role of experts in a process. What is their relation to the stakeholders and how do they contribute their expertise? By way of intermezzo, Sect. 8.3 explores the relation between strategic behaviour on the one hand and substance-driven behaviour on the other hand. Section 8.4 outlines a desirable course of the process from a substantive perspective. It reintroduces and operationalizes the standards of variety and selection.

## 8.2 Bundling and Unbundling of Experts and Stakeholders

It should be recalled that a process approach to decision making is used when a purely substantive approach is impossible. The problems that are to be solved are unstructured, which precludes an unambiguous substantive solution. This causes the need for a process. However, from a substantive point of view, there are two significant risks in this regard.

H. de Bruijn et al., *Process Management*, DOI: 10.1007/978-3-642-13941-3_8,

*Risk 1: Process Displaces Substance—Negotiated Nonsense Rather than Negotiated Knowledge* This risk implies that the parties' interests take such a dominant position in the process that they displace the substantive forces in the process.

For instance, for the sake of consensus, parties accept a process outcome that is appealing to all of them, but that will not measure up to existing scientific insights. 'Anything goes': parties simply decide that a particular problem definition and problem solution are correct, and do not accept any correction by substantive insights or by the views of experts.

If there is an unstructured problem, the parties should, ideally, seek *negotiated knowledge*: substantive knowledge that (1) is accepted by the stakeholders and (2) will bear scientific criticism. If the process displaces the substance, the result tends to be *negotiated nonsense*, which fails to meet the second criterion.

In his study into the realization of large public projects, Robert Bell finds that the construction of these projects is often started before the design has been sufficiently elaborated. He argues that in a context of conflicting interests, a design has two functions. There is a substantive function: a design should direct the building of a project. In addition, there is a process function: a design should serve the interests of the parties involved. The substantive function may suffer if this process interest is too dominant. This may for instance result in designs that are at conflict with the laws of physics, which will obviously cause serious harm to the construction of the project [1].

*Risk 2: insufficient Use of New, Innovative Insights* In addition, there is the risk that the decision making in a process lacks innovation. Too little use is made of new insights, for instance because the managerial parties that participate in the process are simply unaware of these, or because the participating experts do not represent all relevant and available information.

The use of a nuclear weapon results in *blast damage* (destruction as a result of the blast itself) as well as *fire damage* (destruction by the fire storms caused by the weapon). The fire damage is much larger than the blast damage. A 3,000 kiloton bomb hitting the Pentagon will create devastating fires within a 50-mile radius. The lack of this knowledge has had far-reaching implications for the United States' military planning, namely underestimation of the devastating effect of nuclear weapons and, as a result, development of a nuclear weaponry that was much too large. Why did the expert and research communities underestimate the consequences of fire damage [6]?

The explanation is that blast damage belonged to the domain of physicists, while fire damage belonged to that of fire protection engineers (FPEs). The FPEs were less embedded in the academic world, had less computer infrastructure and calculating capacity at their disposal, and therefore had fewer opportunities to model and predict fire damage. As a result, the FPEs were 'not well connected to physicists deeply knowledgeable about nucleair weapon effects' (Ibid., p. 284).

The results of this are as predictable as they are dramatic. The ability of physicists to predict blast damage increases over the years. Researchers focus on their fields of expertise (such as predicting blast damage), and ignore the consequences of fire damage because on the basis of their expertise, it is much more difficult to predict. Eventually this will determine their view on reality: blast damage is important, while fire damage is not. There is cognitive fixation because the community of physicists is too isolated and does not allow any room for the viewpoints of FPEs. These viewpoints are met defensively: whatever does not match the dominant professional belief, is not accepted.

How to deal with these risks [5, p. 100, 10]? The answer is self-evident.

It is important to give knowledge and expertise a position in the process and to secure this position firmly. In this regard it is important to keep in mind that during the past decades, knowledge and science have been generated and organized differently compared to the preceding period. In the world of research and knowledge development, the following developments relevant to process management are ongoing: firstly, science is no longer exclusively directed by its own questions. For some time, science has ceased being the autonomously developing system in an ivory tower. An increasing number of questions that are relevant to science originate from outside of the scientific world. Societal, commercial and political problems play a role in determining which questions are addressed by science.

Secondly, this change has organizational implications. A much larger proportion of research is project-based than before—for instance in public–private partnerships, or in alliances in which knowledge users cooperate closely with knowledge developers.

Thirdly, society is no longer prepared to always wait for science. Society demands answers, even if science is lagging behind. In earlier days, science followed the strategy of seeking publicity only when hard facts were available. Today, however, science is expected to reach increasingly authoritative conclusions before the facts are fully and reliably available. This change has been accurately summarized by Funtowicz and Ravetz [7, 8]. They state the following: while science used to be asked to pass judgment in situations of 'hard facts and soft values', today it is increasingly expected to produce sensible statements about 'soft facts and hard values'. Due to the bundling of science with society, politics and business, science is no longer in a position to be the sole judge of how the generated knowledge measures up. Stakeholders want to have a say in this as well. The result is a process in which a variety of stakeholders evaluates to what extent the generated knowledge is relevant, valid and reliable.

Let us revert to the larger process. Which role in these processes is left for the science that has been generated in this new world, in order to prevent negotiated nonsense and allow innovative insights to enter the process?

## *8.2.1 Four Roles for Experts in the Process*

A common notion in the literature about expert involvement in decision-making processes is that the stakeholders in the process do not accept expert opinions by definition. There are a number of explanations for this—and each of these has its remedy (see Table 8.1).

The first strategy is the most classical one: when an expert's analysis has insufficient authority, the quality of the analysis should be improved. Any opposition of parties against the outcomes of the analyses is countered by improving the

**Table 8.1** Expert involvement in decision-making processes

| | Why are expert opinions not accepted? | Remedy | Relationship substance—process |
|---|---|---|---|
| Strategy 1 | The analysis has insufficient quality | Improve the analysis | Sequential: first the substantive analysis, then the decision-making process |
| Strategy 2 | The stakeholders do not understand the analysis | Improve the communication about the analysis | Sequential: first the substantive analysis, then the decision-making process |
| Strategy 3 | The stakeholders do not commit to the way in which the analysis has been performed, and therefore they do not commit to the result either | Improve the interaction between the experts and the stakeholders about the design and implementation of the analysis, allowing both of them to commit to the results | Sequential: first the substantive analysis, then the decision-making process |
| Strategy 4 | The analysis does not match with the dynamics of the decision-making process | Improve the interaction between the experts and the stakeholders and pay attention to the moment of interaction, allowing the analysis to actually facilitate progress in the process | Analysis and decision making largely proceed in parallel |

analysis and thus by strengthening the authority of the conclusion. This may for instance be accomplished by performing sensitivity analyses for other data or system boundaries. The basic belief is that the expert presents the facts, based upon which the stakeholders in the network will make their decision. 'Speaking Truth to Power', as the saying goes.

The core of the second strategy is communication. Here, too, the basic belief is that the expert presents the facts and that the stakeholders make a decision based on these facts. The latter, however, is not self-evident. There may be significant differences between the language of science and research on the one side and the language of decision making on the other side. Therefore it is important to pay explicit attention to the communication of the results. These should be framed in such a way that they fit into the stakeholders' frames of reference. This is a major theme in risk communication, among other areas. Risk analyses produce results that are difficult to communicate and that therefore fail to have the desired impact. The results of the analyses need to be framed in the language of the decision makers.

The third strategy is focused on interaction. In essence, communication is still a unilateral activity: the point is that experts explain the results of the analyses as well as possible. Interaction, on the other hand, is bilateral: stakeholders are involved in the design of the analysis and in the formulation of its findings [11, 13, 17, 18]. Experts may propose to the stakeholders which data, system boundaries and methodology are to be used. The stakeholders may then seek clarification about this, or they may for instance propose using different data. This may lead to a discussion about the quality of these alternative data, experts may examine the sensitivity of the outcomes to these alternative data, or experts and stakeholders may jointly generate new data. The key idea is that experts and stakeholders arrive at shared views about the analysis method and its results in a process of interaction. This improves the quality of the analysis, since the experts are questioned critically, as well as the acceptance of the results.

Best case is that full consensus evolves. However, there may also be consensus about a number of outcomes while dissensus persists about others. Of course findings that enjoy consensus have more guiding power with regard to the intended decision than outcomes that do not.

Strategy 3 is based on an assumption: once there is agreement between stakeholders and experts about the outcomes of the analysis, these will be used as guidance in the decision making. Experts and stakeholders therefore need to invest in consensus about the analysis prior to the decision-making stage. An important argument can be made against this assumption. A decision-making process has its own dynamics: stakeholders negotiate with each other, they try to raise support for problem definitions and goals, try to conclude package deals, and so on. The agenda is always dynamic, particularly during the initial rounds of a process. This entails a significant risk: experts do not follow the dynamics of the decision making, as a result of which they introduce their substantive insights at the wrong moment: too early, or—more often—too late. Alternatively, their insights pertain to problems that may have been relevant to the process yesterday, but that no longer have any relevance today.

This brings us to a fourth strategy. Experts should follow the dynamics of a process to some extent. Consequently, decision making and analysis do not take place sequentially, but in parallel. This way, experts become a part of the decision-making process.

### 8.2.2 Embedding Experts in the Process

It will be clear that it is the third and fourth strategies that play a role in the processes in this book. The notion that an expert only needs to present a proper analysis (strategy 1) or invest in proper communication of his analysis (strategy 2) does not match with the unstructured nature of problems. Now, how can experts be embedded in a process?

### 8.2.3 *Unbundling of Roles, …*

Experts distinguish themselves from the stakeholders by their expertise, which is less strongly tied to a particular interest. This is an argument for making a clear distinction between experts and stakeholders in a process. Such unbundling implies that the expert and the stakeholder play different roles. The expert can advise the parties and also plays an important role as 'countervailing power' towards the stakeholders. For instance, allowing experts to take a critical look at the draft outcome of a process may prevent negotiated nonsense.

If there is no such unbundling of roles and no clear agreements are made on this point, there is a risk that the expert will become biased towards the interests of one of the stakeholders. In that case, rather than being the person who takes a critical look at the (interim) results of a process or who indicates which innovations are possible, he is someone who justifies decisions by providing relevant substantive argumentation. This risk is particularly imminent in the case of unstructured problems, because these do not have any unambiguous solutions.

### 8.2.4 *… Followed by a Bundling of Activities*

The concept of unbundling of roles calls for an important addition, as is explained in Table 8.2.

Unbundling of experts and decision makers tends to be based on the notion that the expert or researcher discovers the facts, after which the decision maker arrives at a decision—as corresponds with strategies 1 and 2 above. Although such a distinction is obsolete from a science-philosophy point of view, it continues to play a role in the practice of decision making. A logical consequence of such a view is that the roles of experts and decision makers are strictly separated: facts precede judgments. In the case of unstructured problems, however, such unbundling has two major disadvantages:

- The knowledge contributed by experts has no authority for the stakeholders. It is very well imaginable that these stakeholders will not accept the results of a study, for instance, because they disagree with the choice of data, methods or system boundaries. Garbage in, garbage out, as the saying goes. This is why strategy 3 is useful.
- There is a temporal misfit between the expert knowledge and the decision-making process. The results of a study become available either too early or too late, for instance. As a result, science-based criticism may not reach the process in time. Unbundling reinforces this mechanism. This is why strategy 4 is useful.

Research by Jasanoff [12, p. 231] corroborates this. She concludes that processes in which scientific research and decision making are strictly separated stand little chance of authoritative and consolidated decision making. Such decision making is

**Table 8.2** Expert and stakeholder roles, unbundled and bundled

| | Unbundling | Bundling |
|---|---|---|
| Advantages | Expert may act as countervailing power | Influence on the decision making |
| Disadvantages | The expertise is insufficiently authoritative and it is submitted at the wrong time | Expert becomes biased towards certain stakeholders and thus perverts the process |

more likely to take place in processes in which science and decision making are combined.[1]

The Intergovernmental Panel on Climate Change (IPCC) has grown into a relatively authoritative institution, partly because of the way in which expertise and interests are bundled intelligently. The IPCC was established in 1988. It was developed as an intergovernmental body that should assess the existing scientific knowledge on the causes and impacts of climate change, as well as mitigation strategies [16, p. 117]. It is important to note that this IPCC is forced to operate in a strongly politicized environment. The interests are significant and diverse. The uncertainties are significant as well. As a result, the knowledge generated by the IPCC is heavily contested. This became clear for instance in 2010, when some relatively minor errors were identified in the IPCC reports. These minor errors evoked some disproportionately heavy attacks on the IPCC's work. Nevertheless, the impact of the IPCC's conclusions has grown over the years of its existence. It is therefore interesting to examine the institutional way in which expertise and interests have become bundled in the IPCC.

Scientists from the IPCC Bureau are the ones to develop preliminary proposals for the outline of IPCC reports and for the topics of the working groups (ibid.). These scientists are also the ones to write the first drafts of the chapters. The members of the Bureau are chosen by governmental delegates on the basis of nominations from a nomination committee (ibid.). The preliminary concepts of these chapters then enter a review procedure. The first review round is designed like a traditional scientific peer review: authoritative scientists provide commentaries to the first drafts of the chapters. In a second review round, governments are given an opportunity to formulate a reaction. These governments do this by inviting commentary from experts who work at the ministries of these countries, or who work for national research institutes. This governmental commentary should, by the way, be based on published papers in the scientific literature (ibid., p. 118). The governments are in charge of the approval of the summary for policy makers and the synthesis report. Points of discussion regarding the summary are addressed in working groups. In those few cases in which no consensus can be reached about the text of the summary, a dissenting vote will be included in the text naming the dissenter (ibid., p. 120). Countries do not like being singled out in this way, which is why they go through great lengths to reach consensus.

The success and effectiveness of the IPCC is another positive example of bundling of research and decision making. In brief, bundling implies that experts are more aware of the course of the decision making, which is why they are better able to intervene at the right moments. Bundling also implies that experts are better equipped to deal with parties' criticism of their analyses. After all, bundling results

[1] Also see for instance: Tanaka and Hirasawa [19].

in intensive interaction, enabling experts to react properly to parties' criticism. They have more knowledge of parties' views and of the inconsistencies in those views, they have more opportunities for iterations in their research, and so on.

The result is an ambiguous picture: bundling is necessary, but involves the risk that experts become biased towards certain stakeholders; unbundling is therefore desirable, but it involves the risk that experts can play no authoritative role in the decision making. What does this mean for the relation between experts and stakeholders in the process?

- On the one hand, the *roles* of experts and stakeholders should be unbundled (see above).
- On the other hand, it is necessary to bundle the *activities* of the two parties. Starting from their unbundled roles, they should interact intensively in order to avoid the misfits mentioned above. Unbundling prevents experts from becoming biased towards certain stakeholders.
- Bundling of activities may be achieved through a process agreement that stipulates that the stakeholders *have to* submit their (interim) results to the experts at particular moments in the process and *may* submit these (interim) results at other moments.

Such bundling stemming from unbundled roles has two functions: it improves the quality of both the decision making and the knowledge contributed by experts.[2]

### *8.2.5 Improving the Quality of the Decision Making*

Bundling first of all results in an improved quality of the decision making by the parties. After all, bundling forces stakeholders to submit their own views and assumptions to experts. It is not up to these experts to judge which proposals and views are to be chosen (after all, the problems are unstructured, so this would not be possible in an authoritative way); rather, they will indicate how the parties' proposals and views measure up to scientific insights. Their conclusion may be that particular views or assumptions cannot stand the test of scientific criticism.

> The Orange County Landfill Selection Committee has to decide on the location of a landfill. It can choose from seventeen potential locations, which can be compared on the basis of sixteen variables.
>
> The Committee commissions a study, expecting it to provide an objective answer to the question where the landfill should be located. The study, however, produces a different outcome. It shows which judgments can be regarded as objective and where there is room for parties to negotiate. The outcome of the study is that only four of the sixteen variables are relevant when comparing the locations: groundwater, surface water, landfill cover material and isolation. The other twelve variables do not differ between the locations.

---

[2] See for instance: Garvin and Eyles [9].

The study also shows the scores of the various locations with regard to the four variables. One type of conclusion, for instance, could be that establishing the landfill in location X would be particularly harmful to its groundwater.

This way, the study is able to contribute to the substance of the decision making. Parties that invoke one of the twelve non-discriminating variables in the process would seem to have a weak case in the decision-making process. The study also shows where there is room for negotiations because unambiguous conclusions are impossible [14].

The involvement of decision makers reduces the risk of a temporal misfit between the decision making and the research. After all, the decision makers are informed about the current research and because of this bundling, they also have some degree of commitment. Conversely, experts are better able to keep in touch with the decision making and thus are more sensitive to the momentum to submit substantive expertise.

### 8.2.6 *Improving the Quality of the Submitted Knowledge*

Secondly, bundling leads to an improvement of the quality of the research. This is because a critical attitude of the stakeholders vis-à-vis the study and its results will make clear what the underlying values are, which data, methods and system boundaries have been used, which results are robust and authoritative, and which results fail to meet these criteria.

A study by an independent engineering agency shows that the polycarbonate bottle scores poorly in comparison with other packages. One stakeholder in the process, an advocate of the polycarbonate bottle, takes a very critical look at the assumptions upon which the study is based. He makes these assumptions more explicit, and manages to convince the other parties that they are incorrect. This improves the quality of the study into the environmental effects of the polycarbonate bottle.

The fact that every package has its own advocates in the process guarantees that the analyses of the other packages will be examined just as critically [3].

### 8.2.7 *Research and Decision Making: Parallel Connection and Proper Bundling*

Which are the implications of the above for the planning of research activities in a decision-making process? This question is particularly relevant with regard to processes that are highly information-intensive and of which it is clear from the start that a great deal of research has to be done. For instance, decision making about large infrastructural projects will nearly always involve a large amount of research: into the environmental effects, the economic effects, the safety implications, noise nuisance, and so on. The design question is: how should the research be planned in relation to the decision making?

Taking the above into account, it will be clear that a serial planning (research is followed by decision making) is undesirable. Parallel planning is preferable. In this regard, it is important that the research and the decision making should be bundled. The process manager therefore has three tasks:

- He should keep the *roles* of experts and stakeholders separate;
- He should ensure proper bundling of their *activities*;
- He should ensure a *parallel connection* between the research process and the decision-making process.

Once these new tasks have been fulfilled, the process manager can act as an interface between experts and stakeholders. He can use the knowledge of experts to increase the substantive level of the process. We will illustrate this by referring to some examples that we used before.

Situation 1: *Options for improvement* A choice must be made between two options: a carton box and a glass bottle; the question at hand is which of the two has a better environmental profile. Research shows that the carton box has the better profile. Some parties dislike this conclusion. The decision makers then want to know whether the performance of any of these packages can be improved. Researchers calculate the improvement options. Their results indicate that the glass bottle can be improved in several ways, while the environmental profile of the carton box has already been optimized.

The options for improvement might not have been calculated had the experts not played an active role. The decision makers only become interested in the improvement options of the glass bottle after it has become clear that the carton box has a better environmental profile. Had the research and the decision making been connected sequentially, the calculation of the improvement potential might not have been available.

Situation 2: *Standardization* A choice must be made between six packages. Experts announce that four of these packages score poorly, while two score well: the polyethylene bag and the carton box. These are two entirely different packages, whose environmental profiles are determined by entirely different characteristics. With respect to the bag, for instance, loss of product is an important parameter, and the logistics are more complicated: bags require outer packaging, which has its own environmental impact.

The two packages have to be compared, while they are in fact difficult to compare. Researchers therefore apply *standardization*: they use a particular method to make the options comparable. The carton box turns out to be superior to the polyethylene bag.

Without an active role of experts, this standardization would not have taken pace, which would have resulted in a less substantiated choice between box and bag.

Situation 3: *Sensitivity analysis* In a decision-making process about expanding a port area at sea near Rotterdam, a number of stakeholders are opposed to a particular location. The reason is that research shows that expansion in this location would lead to coastal erosion. The supporters of this location, however, argue that the dunes in many seaside resorts are being raised anyway in view of the rising sea level.

Researchers then conduct a sensitivity analysis: how sensitive are the outcomes of the first analysis if an increased elevation of the dunes is taken into account? To what extent will there still be coastal erosion?

This sensitivity analysis would not have been conducted without experts playing an active role, nor would it have been conducted without a parallel connection. After all, the argument of sea-level rise only surfaced during the negotiations.

Situation 4: *Enriching the decision making* In a decision making process, five options are available: options A through E. The decision makers find that a number of combinations are possible: A + B, C + E or D. Research shows that there is a fourth possibility: A + B + C.

This possibility might not have been highlighted without experts playing an active role, and the decision making would have been less rich.

> The process addressing various packages started with the simple question whether disposable or non-disposable packages are better for the environment. Such a dichotomy easily leads to discord and hardly allows coupling between problems and solutions. After all, the parties believe that a choice has to be made between the disposable and the non-disposable package.
>
> The contribution of experts enabled the parties to take a more nuanced view on the problems after the process, resulting in much more opportunities for creative couplings between problems and solutions. Conclusions from the final report by the parties include:
>
> - The dichotomy 'disposable versus non-disposable' is inadequate. There are 'hybrid packages' as well (refill packages, for instance), which are partly disposable and partly non-disposable and have a good environmental profile.
> - Disposable additions, such as caps and clips, have a large impact on the environmental profile of non-disposable packages.
> - The weight of the packages is of major importance; return transport of non-disposable packages within the Netherlands is less relevant.

New options introduced by experts may thus have a positive effect on the process, since they provide new possibilities for negotiation and decision making.

Situation 5: *Room in the decision making* In a decision-making process about the location of wind turbines, six options are available. A dominant party announces that one of these options—a large offshore wind park—is technically not feasible, citing an existing study that corroborates this. The other parties examine this study critically, and then ask the researchers a number of questions. As a result of this interaction, the researchers have to admit that this option is indeed feasible under certain conditions. This room might not have been created without experts playing an active role.

Situation 6: *Unburdening the agenda* Six options are available in a decision making process. The parties involved have to negotiate about these six options. After some time, the parties agree about the variables for which the options will be assessed. There are parties that cling to one particular option out of ideological beliefs. For instance, they are truly convinced that reusable packages are always better for the environment. This seems logical, because some reusable packages may be used up to 40 times. Research shows, however, that for certain reusable packages this line of reasoning is evident nonsense. The parties thus decide not to further consider these options.

Without an active role of experts, this burden would not have been taken off the agenda.

An interesting form of bundling research and decision making is the way in which environmental impact assessments of large-scale physical works are embedded in the decision-making processes about these projects.

For certain projects, environmental impact assessments are compulsory by law. In short, the law obliges a project's initiator to study the project's environmental effects and to report on these in an environmental impact report. The initiator must also develop a number of alternatives to his proposal and describe these, including their environmental impact, in the same report.

Although this law is not very popular among initiators (nor among the competent authorities), environmental impact reports—as a reflection of the research conducted—have proven to strongly affect the thinking and acting of the actors involved. Environmental impact assessments involve the decision makers in the research. It is the initiator who proposes the alternatives and describes the environmental impacts. The competent authority establishes the requirements that the research has to meet. Usually, a specialized research institute conducts the actual study, of course under strict supervision by the commissioning party or the initiator. The scientific quality of the results is evaluated by a committee of independent experts. On the one hand, this committee is independent and may therefore be critical. On the other hand, it is likely to adopt a moderate attitude because it realizes that too much criticism will affect the position of the environmental impact assessment.

In short, although research and decision making are professionally separated, they are in fact bundled. The decision makers leave room to formulate a problem definition, to place certain accents or to emphasize particular conclusions, and there is close contact between researchers and decision makers [20].

## 8.3 Intermezzo: Strategic Behaviour, or Fair and Substantive Behaviour?

Parties behave strategically in complex decision-making processes. For instance, they do not disclose their views because they want to keep their options open, or they take an extreme view to strengthen their negotiating position.

Such strategic behaviour would seem to harm the substance of a process. In determining their position, parties should be guided by substantive arguments rather than by strategic (or worse yet, opportunistic) considerations. This paragraph will, by way of intermezzo, address the distinction between strategic behaviour and behaviour driven by substantive arguments. We will argue that this distinction is problematic.[3]

A distinction between substantive and strategically inspired behaviour matches a particular way of thinking about process management, which may be summarized as follows:

- Good substance is created in a good process.
- A good process implies, among other things, that problem definitions and problem solutions should not be allowed to become fixed too soon. The participants in a process will first have to diverge (i.e., consider a large number of problem definitions and solutions) and should only be allowed to converge at a later stage (i.e., to select one or more problems and solutions). The idea is that the quality of the problem definition and solution is good if these emerge from a wide variety.
- This movement from divergence to convergence tends to be hampered by parties' interests. Interests block the open-mindedness that parties need in order to diverge.
- Inherent to this is the fact that parties will behave strategically when driven by interests. For instance, their hidden agendas will disturb the process of creating good substance. Processes of power will thus corrupt the substantive debate.
- A process of divergence and convergence can only develop in some sort of 'power-free room'.
- Process management is aimed at creating such room. The parties agree, for example, that their debate will be factual and open, that positions in the substantive debate will be valued in the same way, that only the power of arguments will play a role, that they will not play power games, and so on.
- Process management thus comprises two core elements: guiding the substantive debate (with the help of various communication techniques) and—prior to this—making agreements with parties about not adopting strategic behaviour. This will allow the process to take place in the above-mentioned power-free room.

Many types of interactive policy development are based on such assumptions. Substance is good, strategic behaviour is bad, and one of the aims of interactive policy development is not to have substance corrupted by strategic behaviour. Two objections can be raised against this line of reasoning: it is naïve, and it is based on an unjustified distinction between power and substance.

### *8.3.1 Substance Depends on Interest; the Realization of Interests Depends on Strategic Behaviour*

In the case of unstructured problems, various parties may have various legitimate perspectives on problems and solutions.

[3] These sections have been derived from: De Bruijn [2].

Parties' own position and interests play a role in shaping the perspective that they take. When they propagate their perspectives, they will do so partly through substantive argumentation, and partly by strengthening their position in the network—for instance by forming alliances and waiting for the right moment. In other words, they do so by exercising power.

This power exercise is not aimed at corrupting substance, but rather at allowing their legitimate perspective to play a role in the decision making. In a decision-making process, there is competition between various parties with various legitimate problem definitions and solutions. Those who do not join this game may see their own perspective get the worst of it once the decision-making stage is reached. This may affect the quality of the final decision. In other words, the quality of a decision partly depends on strategic behaviour.

### *8.3.2 The Distinction Between Substance and Strategic Behaviour is Debatable*

It is clear from the above that the distinction between substance and power is debatable. Admittedly, power and exercising power may corrupt the substantive debate. However, in a network in which problems are unstructured, the distinction between substance and power is far from unambiguous.

- There is a kind of strategy to promote substantive views in a decision-making process (see above).
- Substantive argumentation can be used strategically, either intentionally or unintentionally.
- The result is a dynamic that is difficult to unravel. When a party in a process changes its viewpoint, is this a substantive enrichment of that view or an opportunistic change, prompted by arguments of power?

In practice, the distinction is sometimes difficult to make, and parties may hold different views about this. One of the experiences with process management is that party A blames party B for strategic behaviour, arousing sincere rage in party B: its behaviour was not strategic, but exclusively inspired by noble, substantive motives [3].

If there is no unambiguous substantive solution, there are by definition different perspectives on the same reality. Consequently, these different perspectives are legitimate, as are the actions to advance this perspective or interest.

We therefore fail to see any valid objections to strategic behaviour, as long as parties' strategic behaviour conforms to certain rules of the game, and as long as parties have a justified interest or a legitimate perspective. The same goes for objections to concepts such as hidden agendas. What is wrong with a hidden agenda if parties know that strategic behaviour is inevitable and may even be legitimate? Hidden agendas are part of the game in a negotiating environment.

**Table 8.3** Two types of process design

| Process design: creates a power-free room | Process design: facilitates negotiations |
|---|---|
| Unbundling of power and substance | Power and substance are inseparable, both factually and normatively |
| Creates a power-free room | Creates a framework for negotiations |
| Offers rules of the game for substantive enrichment | Offers rules of the game for decision making |
| Strategic behaviour is disturbing and therefore undesirable | Strategic behaviour is natural and therefore permitted |

They need not compromise the trust between parties if these parties play the game by its rules.

If parties have to arrive at a decision, strategic behaviour is a given and it is legitimate. A process will therefore have to be designed in a way that leaves room for this strategic behaviour. Something similar applies to process management. Process design and process management must accommodate parties' natural behaviour to a significant degree: the negotiation rules are drawn up with respect for the principle that parties will want to behave strategically.

This process design is the mirror image of the one seeking power-free room (see Table 8.3). In the latter, strategic behaviour is eliminated, and then a substantive debate is held. The opposite of this is a form of process design that regards decision making as a negotiating process and therefore accommodates strategic behaviour. In this respect, too, a process design is contingent and the paradox of change presents itself once again: those who are open to the existing situation may change.

If a process design promises openness and thus bans strategic behaviour, everyone will suspect that there is strategic behaviour anyway. This will generate conflict, since it is contrary to the agreements. If the distinction between substantive argumentation and strategic behaviour is vague, there is a reasonable chance that such suspicions will arise.

However, if a process design permits strategic behaviour—in other words, parties are allowed to show their natural behaviour—such suspicions are far less significant. During the process, trust may evolve between parties, which may moderate their strategic behaviour.

## 8.4 Moving from Substantive Variety to Selection

Unbundling and bundling enable experts to contribute to the substantive quality of decision making. The next question is: from the perspective of the substance and the quality of decision making, what is the most desirable process?

Again, the question about the *substantive* quality of the decision making evokes a *process-based* answer. There is quality if the decision-making process is characterized by *variety and selection*. The idea is that the greater the variety of options discussed, the better the quality of a decision-making process. The quality of the decision-making process will suffer if a particular option is not included in the decision making. Variety is important for the following reasons:[4]

- The greater the variety from which the final option is chosen, the more authoritative this option will be. Its quality would be more debatable if it had defeated only a limited number of other options.
- The actors involved are offered a maximum number of learning processes. The actors participating in the process take note of the options, reflecting on their strengths and weaknesses in interaction.
- In the resulting discussions, parties will explore the possibilities of improving certain options and making them acceptable to a maximum number of actors. This improves the quality of the options.

At some stage in a decision-making process, there will be a transition from generating variety to selecting the best option or options. This selection will have to lead to consolidation: the option or options selected will have to stand up to criticism for some time.

- The first condition for a consolidated selection is that a high variety of options was considered (see above).
- The second condition is that there must be a link with the variety of options considered. There will be no authoritative selection, for instance, if options are selected that were not contemplated in the variety phase.
- The third condition is that the option selected has the support of the parties involved.

The transition from variety to selection also has advantages from a process perspective. If there is sufficient variety at the start of the process, it is more difficult for the parties to submit new ideas in the selection phase. In other words, this strategy simplifies the consolidation of the decision making.

In the earlier-mentioned policy process 'More space for rivers', experts produced a model. This model included 700 potential projects that could contribute to the solution to the problem. For every project, the model presented a variety of data, for instance about the construction costs and the volume of space "gained" for the water. In other words, an enormous variety was created. The transition from variety to selection proceeded relatively fluently in the 'More space for rivers' process. The three requirements were met:

- The list of 700 projects offered sufficient variety. All parties had an opportunity to propose projects.
- The final selection was based upon this variety. Some options were in fact added along the way, but these enjoyed consensus.
- Parties agreed, *grosso modo*, about the selection of 40 projects.

Which are the implications of this initial variety? Firstly: there are probably no additional options, which is conducive to the quality of the decision making. Secondly, in an environment with this much variety, it is easier for parties to learn. A party will realize, for

4 See for instance: de Jong [4].

instance, that by deepening its port it will kill two birds with one stone: there will be more space for the river as well as a better accessibility for larger ships. This may inspire this party to make such a coupling. Thirdly, the large amount of options allows parties to play with these options: they can exchange them, couple them, seek alternatives for unacceptable options, and so on. In other words, the options are enriched. These are all substantive advantages, but there is a process-related advantage as well. If parties participate in this process and there is a final result, it is difficult for parties to ignore this result based on the argument that other options are available. After all, all available options played a role in the process.

### *8.4.1 The Transition from Variety to Selection*

The advantages outlined above may be clear, but the criterion of variety and selection is still difficult to operationalize. A number of questions need to be answered if the requirement of variety and selection is to provide any support. For instance: how serious should an option be in order to be taken into consideration? At which moment should the options be known? Who decides which option will be taken seriously? What does 'taking into consideration' actually mean?

A process manager who is wondering about the chance that new, important options will emerge during a particular process, can get some idea of this by checking whether the decision-making process is continuing to offer learning processes to the parties involved. The associated operationalization is therefore: It is fruitful to continue the variety-generating phase as long as the actors participating in the process continue to learn.

Two types of learning may be distinguished: cognitive learning and social learning.

Cognitive learning involves the question whether parties still produce new facts, views, values, arguments, thinking patterns, and so on. As long as the process manager notes that this is still the case, there is a chance of new variety being created. If the parties go on reproducing the same facts, views, and so on, there is no longer any cognitive learning.

Social learning involves the question whether parties still establish new relations and interactions. These may result in new insights and therefore new variety. All of this means that many transitions from variety to selection do not proceed smoothly, but seemingly chaotically. Although this may seem to be a sign of a bad process, it may actually indicate the quality of a process[5] (see Fig. 8.1).

A process is ripe for selection if there is cognitive and social stabilization.

Selection takes place at moments 1, 2 and 5. The figure also shows that the transition from variety to selection is marked by iterations: moments 3 and 4. Sometimes selection has taken place, but participants go back on this, for instance because they have established new relations. Of course, the process will be

---

[5] See for instance: Salsich [15].

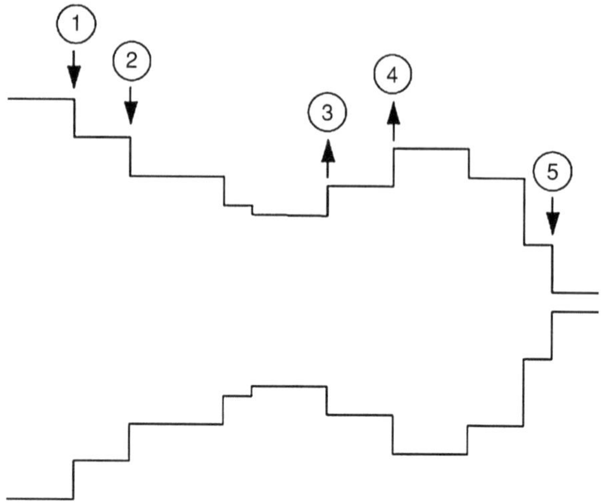


**Fig. 8.1** A transition from variety to selection

jeopardized if this happens too often. This is another argument for not concluding the variety-generating phase too early.

As mentioned before, when parties are offered room for iterations, it is easier for them to participate in the process wholeheartedly. It is up to the process manager to make all effort to prevent the parties from using this room.

Table 8.4 summarizes the argumentation of the preceding paragraphs. There are two arrangements that may safeguard the substantive quality of decision making in a process: (1) unbundling of the roles and bundling of the activities of experts and stakeholders, and (2) a transition from variety to selection. Table 8.4 shows the consequences for the substance and the quality of decision making. In addition, it illustrates that these arrangements also have positive effects from a process perspective.

## *8.4.2 The Substantive Expertise of the Process Manager*

The above shows that the balance between substance and process is a major concern for the process manager. He must prevent the substance from displacing the process and vice versa. In the end, this prompts the question which substantive expertise the process manager should have.

Earlier, we described the risks associated with a model in which the process manager also acts as a substance expert. The process manager's interference with the substantive course of events may then be strong enough to put pressure on the process aspects.

One may wonder, however, which substantive knowledge the process manager should have himself. Even if the process manager does not play the role of substance expert, he must have substantive knowledge. If he does not, he risks not

**Table 8.4** Summary of the argumentation in Sect. 8.4

| | Consequences for the substance and quality of the decision making | Consequences for the process |
|---|---|---|
| Experts and stakeholders: unbundling of roles, bundling of activities | A sense of the capriciousness of decision-making process | Efficiency: particular options are excluded from the decision making. |
| | More authority of research findings | Room: new options may simplify decision making |
| | Expert is not influenced by parties' interests; no bias | |
| Transition from variety to selection | All possible options have been taken into account | Consolidated decision making: it becomes more difficult to ignore the decision making |
| | Learning processes | |
| | Selection is more authoritative | |

being taken seriously by the parties. However, a process manager who is strongly guided by substantive knowledge runs the risk that the substance will displace the process, which will also harm his position.

This tension may be relieved by distinguishing between substantive knowledge of the first, second and third order:

- Substantive knowledge of the first order implies that a process manager is able to ask the right questions;
- Substantive knowledge of the second order implies that the process manager is able to evaluate the answers to these questions adequately;
- Substantive knowledge of the third order implies that the process manager himself is, or would be, able to answer these questions.

At the very least, the process manager should have substantive knowledge of the first order. A good process manager will usually manage to acquire substantive knowledge of the second order during the process. The risk of substantive knowledge of the third order is that the process manager will interfere too much with the substance of the process, which will harm his position.

## References


1. Bell R (1998) The Bottomless Pit: megaproiects and manipulation. Translated into Dutch by H. Vander Kooy. Aristos, Rotterdam
2. De Bruijn JA (2000) Processen van verandering. Lemma, Utrecht
3. De Bruijn JA, ten Heuvelhof EF, in't Veld RJ (1998) Procesmanagement: Besluitvorming over de milieu- en economische aspecten van verpakkingen voor consumentenprodukten, Delft, commissioned by Foundation Verpakking en Milieu

4. De Jong WM (1999) Institutional transplantation: how to adopt good transport infrastructure decision making ideas from other countries? Eburon, Delft
5. Dutch Scientific Council for Government Policy (2008) Onzekere Veiligheid. Amsterdam University Press, Amsterdam
6. Eden L (2004) Whole World on fire: organizations, knowledge, and nuclear weapon devastation. Cornell University Press, Ithaca
7. Funtowicz SO, Ravetz JR (1992) Risk management as a postnormal science. Risk Anal 12(1):95–97
8. Funtowicz SO, Ravetz JR (1993) Science for the post-normal age. Futures 25(7):735–755
9. Garvin T, Eyles J (1997) The sun safety metanarrative: translating science into public health discourse. Policy Sci 30(2):47–70
10. Gibbons M, Limoges C, Nowotny H, Schwartzman S, Scott P, Trow M (1994) The new production of knowledge: the dynamics of science and research in contemporary societies. Sage, London
11. Giddens A (1994) Beyond left and right: the future of radical politics. Stanford University Press, Stanford
12. Jasanoff S (1990) The fifth branch: science advices as policy managers. Harvard University Press, Cambridge
13. Mayer IS (1997) Debating technologies. A methodological contribution to the design and evaluation of participatory policy analysis. Tilburg University Press, Tilburg
14. Miranda ML et al (1996) Informing policymakers and the public in landfill siting processes. In: Technical expertise and public decisions. Institute of Electrical and Electronic Engineers, Princeton
15. Salsich PW (2000) Grassroots consensus building and collaborative planning. Festschrift 3:709–740
16. Siebenhüner B (2003) The changing role of nation states in international environmental assessments-the case of the IPCC. Glob Environ Chang 13(2):113–123
17. Steelman TA (1999) The public comment process: what do citizens contribute to national forest management? J For 97(1):22–26
18. Steelman TA (2005) Elite and participatory policymaking: finding balance in a case of national forest planning. Policy Stud J 29(1):71–89
19. Tanaka Y, Hirasawa R (1996) Features of policy-making processes in Japan's council for science and technology. Res Policy 25(7):999–1011
20. Ten Heuvelhof EF, Nauta C (1997) Environmental impact: the effects of environmental impact assessment in the Netherlands. Project Appraisal 12(1):25–30

# Chapter 9
# A Concluding Remark

Over the past years, we have had many discussions about process management—with initiators of large-scale projects, with governments that were facing resistance, with organisations that had doubts about participating in a particular process, with project managers who noticed that their project was stagnating. Many of the notions in the preceding chapters surfaced during those discussions.

In addition, we have reflected a great deal on this kind of processes together with politicians and managers, and during presentations, master classes and intercompany trajectories. An important question that was often raised in the context of these reflections relates to the tension between horizontal and vertical management. To conclude this book, we would like to make some remarks about this tension.

Vertical management occurs via the formal, hierarchical lines. The director of a business unit is managed by a member of the Board of Managers. A public official is managed by a director, who, in turn, is managed by a minister. Horizontal management, on the other hand, refers to processes of the kind that is described in this book. The director of a business unit negotiates with his clients and with his colleagues/competitors. The public official consults with societal players about a new project.

There may be a tension between these two lines. Simply put, the parties in the process may, perhaps after a lengthy negotiation, have a certain preference that is not acceptable to the actor who is superior in the hierarchy. How to deal with this tension?

Suppose a public official is negotiating on behalf of a minister in a complicated process about road pricing. This process involves civil society, regional and local authorities, business and trade organizations and scientists. With much effort they manage to conclude a number of agreements, including on road pricing. The result is then submitted to Parliament, which disapproves of it and amends it to such an extent that the entire agreement package falls apart.

An important observation has to be that the tension between horizontal and vertical is simply a given—and that it is no use to deny it or to regard any one of these types of management as superior. Sometimes it is quite frustrating to parties

H. de Bruijn et al., *Process Management*, DOI: 10.1007/978-3-642-13941-3_9, 

in a process when agreements that have been concluded in horizontal consultations are simply dismissed by the body that is superior in the hierarchy. This may create the impression that a government is unreliable: agreements with its representatives and investment in a lengthy process prove to be of little value.

However, the tension between vertical and horizontal may also be frustrating for the superior actor. A parliament, for instance, may be presented with a complicated package deal for its evaluation. It may either adopt or reject the package. Rejection entails a large number of negative consequences: the entire package deal falls apart and the process has to start all over again—if the other parties even wanted to go along with that. Therefore it is common for parliaments to have no actual freedom of choice. This may create the impression that democratic control is not functioning properly, particularly when a process was not really open, but rather had an over-representation of certain actors—whether 'old boys' or activist one-issue groups.

This tension, again, is a given. When vertical prevails over horizontal, this may be frustrating—but the opposite may be equally frustrating. Which arrangements can help relieve this tension?

- Firstly, room is created if this tension is acknowledged. When hierarchically superior actors acknowledge that useful horizontal processes exist as well, they may offer their subordinates room to participate in these processes. When parties in the process accept that there is also a vertical line, they may offer each other room as well—after all, every party needs to account for the process results in its own vertical line. An important strategy for players in a process is therefore that they make this tension explicit, and that they communicate in horizontal as well as vertical processes.
- Secondly, for the players in the horizontal process it is important to realize that the chances of acceptance in the vertical line increase along with the quality of the horizontal process. Quality, in this case, does not only apply to the substance of the final result, but also to the process design. Have a sufficient number of actors been involved in the decision making? Has there been a sufficiently substantive variety of ideas? The more parties and the more substantive variety in a process, the more difficult it is for the superior actor to ignore the results of the process. After all, everything has been discussed, and everyone has been part of the process.
- Thirdly, respect for the players' core values is important in this regard. In the example of the parliament that is faced with a fully negotiated package deal, the political primacy is the core value. In a democracy, the parliament has the first and the last say, and this principle ought not to be looked upon with disregard. If there is no respect for this core value, and if this parliament is put under constant pressure to live with the package deal, this parliament will not be inclined to participate in this process. If there is respect, however, participation in the process is more appealing, and views held by the process participants might actually enter the vertical line.
- Fourthly, the participants in the process may agree upon some kind of 'comply or explain' arrangement. Every participant in the process is offered an

opportunity to deviate from the agreements if he or she is unable to get these agreements accepted in the vertical line. If this is the case, and there is deviation from the agreements, this has to be justified vis-à-vis the other process participants. This is some incentive not to use the possibility to deviate from the agreements too lightly.
- Fifthly, parties in a process may conclude agreements on how to deal with vertical interventions. Such an agreement may for instance stipulate that the superior actors are granted some room to amend the adopted agreements. Alternatively, parties may agree to continue their negotiations if a superior actor has amended or rejected certain agreements.

There is no doubt that there are other strategies imaginable. Our point is that each of these strategies is based upon the idea that there is a tension between vertical and horizontal, and that arrangements are therefore needed to deal with this. In our opinion, designing such arrangements is more fruitful than attempting to create harmony between vertical and horizontal structures—which will never succeed—or regarding any of these kinds of management as superior—as there are valid arguments for both of them. A good process architect and process manager will constantly be aware of this tension and create such arrangements. Moreover, while the tension between vertical and horizontal may be frustrating, it may also be fruitful for a superior body to take a critical look at the agreements that were concluded in a process. After all, process logic may be overly dominant in a process. We have witnessed this all too often: parties negotiate in a process, the composition of parties is often quite random, the exit option is too costly for these parties, so they will need to reach consensus at some point, even if this results in substantive outcomes that are undesirable.

The following quotes by Henry Kissinger and Friedrich Engels reflect this quite nicely:

> The alternative to the status quo is the prospect of repeating the whole anguishing process of arriving at decisions. This explains to some extent the curious phenomenon that decisions taken with enormous doubt and perhaps with a close division become practically sacrosanct, once accepted.

> Denn was jeder einzelne will, das wird von jedem anderen verhindert, und was herauskommt, ist etwas, das keiner gewollt hat.

> [For what is preferred by any individual is rejected by all of the others, and what finally results is something that nobody wanted.]

Thus, we conclude with two critical quotes about processes. To some readers this book may be like a handbook: it offers some suggestions on how to design and manage processes. However, processes are shaped by intelligent parties, and surprises therefore always lurk around the corner. In addition, processes may become a goal in themselves when the quest for consensus overrules various other considerations. Process architects and process managers therefore need to be constantly critical towards the processes that they might have designed or managed themselves.

# Index

MIX
Papier aus verantwortungsvollen Quellen
Paper from responsible sources
FSC® C105338

If you have any concerns about our products,
you can contact us on
**ProductSafety@springernature.com**

In case Publisher is established outside the EU,
the EU authorized representative is:
**Springer Nature Customer Service Center GmbH**
**Europaplatz 3, 69115 Heidelberg, Germany**

Printed by Libri Plureos GmbH
in Hamburg, Germany